D1254834

MANIPULATED
SCIENCE

Books by Mark Popovsky
In Russian

THE WAY TO THE HEART (1960)
THE STORY OF DR. HAFFKINE (1963)
THE THOUSAND DAYS OF ACADEMICIAN NIKOLAI VAVILOV (1966)
THE MAP OF HUMAN SUFFERING (1971)
PANACEA, DAUGHTER OF AESCULAPIUS (1973)

MANIPULATED SCIENCE

*The Crisis of Science
and Scientists in the
Soviet Union Today*

Mark Popovsky

translated from the Russian by Paul S. Falla

DOUBLEDAY & COMPANY, INC., GARDEN CITY, NEW YORK, 1979

ISBN: 0-385-14495-4
Library of Congress Catalog Card Number 78-14708
Translation Copyright © 1979 by Paul Falla
All Rights Reserved
Printed in the United States of America
First Edition

DEDICATED WITH AFFECTION AND GRATITUDE TO
TATYANA VELIKANOVA, MATHEMATICIAN, AND
IGOR KHOKHLUSHKIN, ECONOMIST

CONTENTS

ILLUSTRATIONS

follow page 102

AUTHOR'S NOTE

My first literary teacher was Paul Henry de Kruif, whose eloquence imparted a warm human atmosphere to chilly schoolrooms and made the study of science a personal and almost intimate affair. When I grew older and my comrades were immersed in books about explorers and revolutionaries, I lived in the world of Pasteur and Koch, Mechnikov, Behring, and Roux. I did not think of these men as Olympians: De Kruif had not been afraid to point out strange and humorous aspects of the lives of world-famous scientists. He felt on equal terms with the great ones, paying due tribute to their genius, but not forgetting their oddities. This free-and-easy manner impressed me; and even now, when I have written a score of biographies of Russian physicians and biologists, I cannot help feeling somewhat envious of my American colleague—not so much for his talent, which after all was innate, but for the lack of embarrassment with which he treated his subjects. To him, a scientist, however eminent, was first and foremost a human being, and he made no secret of his attitude toward the great man's personality. I, alas, have seldom been allowed to imitate this. Soviet editors and censors prefer Russian scholars to be surrounded with an aura of perpetual triumph: it was wrong, they admonished me, to make jokes about these great men—still more to point out their mistakes.

Nonetheless, I had the satisfaction of spending many years in the company of outstanding workers in the field of science, which seemed to me a sanctuary of everything that was best in human achievement. I came to love the scientists I wrote about, and they deserved no less: Vladimir Haffkine (1860–1930), for instance, who conquered plague and cholera and saved millions of lives, or Nikolai Vavilov (1887–1943), who waged war on famine and traveled throughout the world in search of cultivated plants.

Living heroes were in no way inferior to their great predecessors. How could one fail to pay homage to Mikhail Hadjinoff (born 1899), a biologist from the provinces who, despite the risk to himself and his family, carried on genetic experiments for many years on lines that were forbidden by the Lysenko dictatorship. These experiments led to outstanding discoveries which would never have been achieved but for Hadjinoff's immense courage. Another hero of mine, Valentin Voyno-Yasenetsky (1877–1961), a brilliant surgeon and pioneer of the surgery of sepsis, took holy orders during the Soviet period and became a bishop under the name of Luke. He spent twelve years in prison and exile simply because he refused to give up his faith or to abandon the pursuit of science. What characters and what personalities were these, and what an example to future generations!

During the years when I was studying the lives of these great men, the number of scientists in the Soviet Union increased tenfold. The old generation died out and was replaced by teams of "scientific workers," organized in large institutions to solve scientific problems by sheer force of numbers. The typical scientist was a mass-produced operator, one of a crowd; the importance of personality dwindled to the vanishing point. This new, anonymous band of scientists bore no resemblance, either spiritual or social, to the world described by Paul de Kruif. I went on looking for heroes about whom I could write books, but the search became more difficult year by year. Increasingly I was forced to the conclusion that the Soviet scientists of the 1970s were scarcely to be recognized as successors of Pasteur, Darwin, Rutherford, and Pierre Curie: both in psychology and in ethics, they were faceless products of mass education. As a biographer in search of well-defined personalities, I felt out of place amid the throng of identical scientists, none of whom deserved the proud title of a "man of science." It may be that this state of things reflects a trend common to the whole civilized world in the second half of the twentieth century. Norbert Wiener, Albert Schweitzer, and others have drawn attention to the erosion of personality and the lowering of ethical standards in the laboratories of Europe and America. That, however, was not my problem: I felt that my occupation in Russia had gone, that I had nothing and nobody left to write about.

Then I remembered my diaries. I had kept them for many years, taken them about the country with me and recorded impressions that were not for other eyes and that could not be published under Soviet conditions. My notes reflected an aspect of reality that Soviet writers are not allowed to describe: the truth about the new type of science and the new breed of scientists. I read through what I had written, consulted letters and archives, and decided to write a new kind of book, for which Paul de Kruif could provide no inspiration: an account of "Soviet scientific man," the mass-produced directors of institutions and laboratories throughout the USSR. It may indeed be doubted whether they truly deserve to be called scientists, but I decided not to dispute the official term. Accepting the definition of a "scientist" as "anyone occupying a responsible post in a scientific institution," I have titled my first chapter "The Million"—for, according to the latest Soviet statistics, the number of scientists working in the Soviet Union is no less than 1,200,000.

Mark Popovsky
December 1978

"Science lies in the hands of
Soviet state and is sheltered by
their protecting warmth."

Academician L. A. Artsimovich (*1909–1973*)

MANIPULATED
SCIENCE

CHAPTER 1

THE MILLION

"Of course, with a supply like this one cannot count on much that is positive or creative, but it comes in very handy as a weapon and a means of impeding the course of society."*

In 1975, on the solemn occasion of the two hundred fiftieth anniversary of the Academy of Sciences of the USSR (formerly called the Imperial and then the Russian Academy), the publishing house Nauka produced a complete list of all past and present members of this institution, including honorary and corresponding members and foreigners. The list consisted of two substantial and handsome volumes, printed on excellent paper, in a light brown leather binding stamped in gold. What could be a better testimonial, to commemorate for posterity the names of all those who, during the past two and a half centuries, had borne the proud title of Russian academicians?

Turning over the glossy pages, one was at first impressed to see the faces of generations of immortals, reproduced in clear-cut miniature portraits.

But suddenly, on coming to the Soviet period, one becomes aware of a strange editorial lapse: the dates of the deaths of certain academicians are given with a mention of the year, but not of the month or day. Sometimes even the year is indicated as doubtful.

The physicist and chemist E. I. Shpitalsky died as recently as 1931, but the editors of the commemorative volumes did not seem to know the month or the day; and the same was true of three well-known historians: K. V. Kharlampovich (1932), Yu. M. Sibirtsev (1933), and S. V. Rozhdestvensky (1934). Similarly it is specified in the volumes, without further precision, that the histo-

* M. E. Saltykov-Shchedrin, *Complete Works* (in Russian), 1965 ff., Vol. 9, p. 37.

rian V. G. Druzhinin and the philologist G. A. Ilyinsky died some time in 1937, while the agriculturist N. M. Tulaykov, who was by no means an old man, for some reason disappeared in 1938. Another philologist, G. F. Tsereteli, died "on or after March 24, 1938," and the engineering expert V. Yu. Gan "in or after 1939." G. A. Levitsky, a cytologist of world reputation, is said to have departed this life "in or after July 1945," and the astronomer K. D. Pokrovsky "on or before November 17, 1945." Altogether the compilation contains a great many vague and incomplete references of this kind.

Such forgetfulness is, to say the least, unexpected—after all, the volumes deal with prominent scientists of our own day, and not of the remote eighteenth century. But there are more surprises to come.

Many scholars who are well known for having belonged to the Academy are simply not mentioned at all in the pages of the commemorative volumes. It is perhaps not wholly surprising that the compilers left out political figures such as Stalin's rival N. I. Bukharin (1888–1938), or even D. B. Ryazanov-Goldendakh (1870–1938), the founder and first director of the Marx-Engels-Lenin-Stalin Institute. But in addition to these absences, we look in vain, for instance, for the name of Ivan Alekseyevich Bunin (1870–1953), a winner of the Nobel prize for literature who was elected an honorary academician in November 1909. What has become of him? Also, we fail to find two world-famous organic chemists, Vladimir Nikolayevich Ipatieff (1867–1952) and Aleksei Evgenyevich Chichibabin (1871–1945); yet both were full members of the Academy and founders of important scientific schools. Nor do we find such celebrities as the economist and statistician Petr Berngardovich Struve (1870–1944), the historian Mikhail Ivanovich Rostovtsev (1870–1952), or another historian, Aleksandr Aleksandrovich Kizeveter (1866–1933). The physicist and biologist Georgi Antonovich Gamow (1904–1968) is not there, although he was, after all, quite an important man: he was the first to discover the genetic code, he propounded the "big bang" theory of the universe, and received a United Nations prize for his work in popularizing science. Yes, there was an academician of that name, but he has somehow been overlooked—his name doesn't figure in the list. . . .

Let us lay aside the handsome jubilee volumes and reflect on the compilers' strange "mistakes." The careful reader will of course be well aware by now that the editor in chief, G. K. Skryabin, academic secretary to the USSR Academy of Sciences, knew precisely what he was doing. He and his assistants were, in fact, perpetrating an overt forgery, or rather a series of forgeries—there is no space to go into them all—with the object of concealing the truth about what happened to the Academy after 1917—the arrests and executions, the flight of scientists abroad, the wholesale destruction of laboratories and institutes. Let there be no mistake about it: sooner or later the Soviet public will know about all these things.[1] Meanwhile, the compilers of the two handsome volumes have done us a service: without intending it, they have revealed the fact that since 1917 many of the leading lights of Russian science have been persecuted, deprived of a livelihood, and forced to emigrate to all corners of the earth. It is true that these volumes refer to academicians only, but there are enough documents and witnesses to confirm that, from the October Revolution onward, the Soviet government consciously made war on the Russian scientific intelligentsia as a whole.

At the turn of the century, the intellectual community in Russia was not particularly large. Even in 1914, people in learned professions, including physicians, conservatory professors, and teachers of theology, numbered no more than 11,600—a far smaller total than Sweden could boast of at that time, let alone Britain, France, or Germany. Moreover, the educated population was very unevenly distributed around the country. A modest number of academics were to be found in Moscow, Warsaw, Kazan, Kiev, Odessa, and Tomsk; but the vast majority preferred to live in St. Petersburg, the capital city which housed the Imperial Academy of Sciences, the Institute of Experimental Medicine, the University, the Academies of Military Medicine and Forestry, the Institutions of Higher Education for Women, and other learned bodies and research institutes. After the Revolution the first onslaught on Russian science by the new regime took place in Petrograd (later Leningrad). Its first victim was the Imperial Academy of Sciences.

The downfall of the monarchy in February 1917 was greeted with joy by the Academy. In a special session on March 24, the members "unanimously resolved to make available to the Government, which enjoyed the people's confidence, the knowledge and resources with which the Academy might be of service to its country."[2] They took a different view, however, of those who stormed the Winter Palace eight months later. As the great physiologist Ivan Pavlov put it in the winter of 1918, "If what the Bolsheviks are doing to Russia is an experiment, it is one to which I would not subject even a frog."[3] Most of his fellow academicians were of the same opinion. They set great store by the democratic hopes aroused in February, and favored the convocation of the Constituent Assembly—for which they duly paid after it was dissolved by the Bolsheviks in January 1918.

The Soviet leaders resolved to starve the scientists into submission. On April 12, 1918, the Council of People's Commissars, chaired by Lenin, decreed that in the future, financial aid would be granted to the Academy only for "appropriate projects"—in other words, for such economic and technical research, of very limited scope, as the government might choose to commission. Apart from this, no provision whatever was made for the maintenance of the Academy's institutes, museums, and laboratories.

The Academy continued to fight for its independence, for the "neutrality" of science and its freedom from party control. It suffered for these efforts. In 1932 a man named Orlov described with malicious pleasure the fate of Russia's chief intellectual center:

> For the first seven or eight years the Academy tried to go on functioning as before. . . . It turned in upon itself. . . . Its connections at home and abroad were restricted. . . . The intake of new material dwindled to a minimum. . . . The scale of research work diminished. . . . Year by year the Academy was forced to reduce its output. More and more material was prepared for publication but not printed. The flood of 1924 did serious damage to its collections.[4]

Caught in a financial vise and at times almost reduced to beggary, the Academy nevertheless held out for several years. It produced "nonpolitical" works and accepted into its ranks historians and philosophers who were far from believers of dialectical materialism. It not only maintained institutes of physics, chemistry, and zoology, but—incongruous as this was under Soviet conditions—continued to run an institute for the study of Buddhist culture and a commission charged with producing a scholarly edition of the Bible.

However, while the Academy as an institution managed to defend itself for some time, many of its members had a hard fight against cold and starvation. They struggled as best they could, finding solace in their work. Ivan Pavlov continued his experiments, using stray dogs which his assistants caught in the streets. Nikolai Kravkov, the pharmacologist, took advantage of the perpetual cold in his laboratory to demonstrate the effect of freezing on live tissue. Orest Khvolson, a celebrated physicist, wearing a winter overcoat, boots with galoshes, and cotton gloves, sat at his desk in two degrees Celsius and worked at a book on the importance of physics to mankind.

But cold and hunger finally got the better of the Petrograd intellectuals. One of the first to succumb was Professor I. A. Velyaminov, founder of the school of field surgery. A few days before his death he attended a meeting of the Pirogov Society and, facing a bust of the famous surgeon, took off his academic cap and solemnly exclaimed: *"Ave Pirogov! Morituri te salutant!"* This allusion to the doomed gladiators was highly appropriate; one after the other, the following eminent scholars died in their freezing apartments:

the brilliant linguist and academician Aleksei Aleksandrovich Shakhmatov;

the outstanding mathematician Andrei Andreyevich Markov;

the academician Yevgraf Stepanovich Fedorov, professor at the Mining Institute and formulator of the theoretical basis of crystallography;

Aleksandr Aleksandrovich Inostrantsev, a corresponding member of the Academy, who owned a unique geological collection and taught more than one generation of geologists;

the botanist Christopher Gobi, the zoologist Valentin Bianki, the geologist Petr Kazansky—and many others.

Later Soviet historians blamed the mortality among scientists in Petrograd on the Civil War, the blockade, and foreign intervention. But Yudenich's two attempts to capture the city lasted only a matter of months—from May to November 1919—and a year later, in the fall of 1920, Professor (later academician) Nikolai Vavilov wrote: "The ranks of Soviet scientists are getting thinner day by day, and one trembles for the fate of learning in our country. For many are called, but few are chosen."[5]

Another year passed, but still no one thought of relieving the sufferings of the Petrograd scholars: deprived of grants and food rations, they continued to die. In March 1921, Maksim Gorky wrote to H. G. Wells: "There is no food, and it is not an exaggeration to say that there will soon be a famine in Petrograd. . . . I simply cannot imagine how our scientists will manage to live through the coming weeks."[6] The next year, in May 1922, Nikolai Vavilov, director of the Botanical Institute, wrote officially to the Commissariat for Agriculture: "We are receiving no funds either to pay our staff or for operational expenses. . . . Even though all the staff are ready to put up with hardship and to be satisfied with the absolute minimum, the position is really impossible. Food coupons arrive a whole month late. . . . The staff have not been paid for two months. . . ."[7]

The following year, 1923, conditions were no less inhuman; in a letter to a botanical colleague, Vavilov wrote: "It's very difficult to exist here . . . things are nearly as bad as in 1919."[8]

Scientists outside Petrograd were no better off.

Professor Mikhail Semenovich Tsvet, a botanist and the creator of chromatographic analysis, died on June 26, 1919, in Voronezh: he had lost all his possessions and was in abject poverty.

Mikhail Kurako died of typhus at Kuznetsk in February 1920, shortly after completing a major work on the construction of blast furnaces. In the general climate of lawlessness and open hostility toward intellectuals on the part of the regime, his apartment was pillaged and the manuscript destroyed.

The list of martyrs among Russian scientists during the 1920s could be extended indefinitely. Those who were not killed by cold and hunger, typhus and typhoid, were finished off by the authorities, who regarded science with suspicion when they tolerated it at all. Scientists were mostly of nonproletarian origin, and who knew what they were up to in their laboratories? It was safer to put them behind bars, or beat them to death, or rob them, or at any rate stop them from working. In the *Bulletin of the Russian Academy of Sciences* at this period, we read: "The Permanent Secretary reported that Professor V. A. Zhukovsky, a corresponding member, had been expelled from his post by force."[9] Zhukovsky occupied the chair of Persian literature at Petrograd University, and the force used was such that the sixty-year-old professor died a few days later, on January 4, 1918.

This was only the beginning, but it was a significant one. In August 1921 the Petrograd secret police, CHEKA (Chrezvichainaya Komissiya), executed Professor V. N. Tagantsev, a geographer, with sixty other members of the so-called Tagantsev Conspiracy: these included the poet Nikolai Stepanovich Gumilev and the sculptor Ukhtomsky, the geologist Kozlovsky, the technologist Professor Tikhvinsky, and Professor Lazarev, an expert in political science.

The Tagantsev affair was only one of many fabrications for the purpose of victimizing scientists. The "security organs" were active in Petrograd and Moscow, and especially so in the reconquered Crimea. After the outbreak of the Civil War, many of the country's chief scholars had taken refuge at Simferopol. From 1918 to 1922 students at Taurida University could hear lectures by the geochemist V. I. Vernadsky, the biochemist A. V. Palladin, the mathematician I. M. Krylov, the physicist I. E. Tamm, and the philologist N. K. Gudzy. Another teacher at the university was Yakov Ilyich Frenkel, a future corresponding member of the Soviet Academy. At Simferopol, Frenkel was considered a "Red": he was put in prison by Denikin,[10] where he welcomed the arrival of the Bolsheviks. But as soon as they took over, the "special section" of the Black Sea Fleet started to arrest scholars right and left. Taurida University was disbanded, and in January 1921 Frenkel left for Moscow, bearing an extensive report on conditions in the Crimea which he intended to show to Anatoli Vasilyevich Lunacharsky, Commissar for Education and Culture. In it he wrote that "most of the bitterest adversaries of Soviet power have left the Crimea. The continued terror is turning neutrals and even sympathizers into enemies." M. N. Pokrovsky, Lunacharsky's deputy, showed the political section of Frenkel's report to Lenin,[11] but no change took place in the Crimea, where the excesses were due not to local authorities getting out of hand but to directions from above.

As early as the fall of 1919, Lenin wrote to Gorky, who had protested the lawless treatment of intellectuals in Petrograd:

You utter incredibly angry words about what? About a few dozen (or perhaps even a few hundred) Cadet [Constitutional Democrat] and near-Cadet gentry spending a few days in jail. . . . A calamity, indeed! What injustice! A few days, or even weeks, in jail for intellectuals. . . . We *know* that the near-Cadet professors quite often *help* the plotters. That's a fact.

The intellectual forces of the workers and peasants are growing and gaining strength in the struggle to overthrow the bourgeoisie and its henchmen, the intellectual lackeys of capital, who imagine they are the brains of the nation. Actually, they are not the brains, but shit.[12]

A few days later Lenin repeated the same idea in a letter to Gorky's wife, M. F. Andreyeva: "To prevent conspiracies we must not refrain from arresting the *whole* Cadet and near-Cadet fraternity. They are all capable of helping the plotters. It is a crime not to arrest them."[13]

Lenin, with his fanatically partisan approach, regarded the whole intellectual community as nothing but "Cadets" and potential enemies; and he argued that because they *might* on occasion help the conspirators, it was only proper to arrest them without delay. He would have liked to shoot the whole rabble without further ado, but unfortunately specialists and technicians were needed to increase the "productive forces of the republic." Lenin's solution to this dilemma was expounded in his speech to the Eighth Congress of the Russian Communist Party in March 1919, in which he impressed on local authorities that they must keep a close watch on the "bourgeois experts," placing them "in an environment of comradely collaboration, of worker commissars and of communist nuclei; they must be so placed that they cannot break out; but they must be given the opportunity of working in better conditions than they did under capitalism, since this group of people, which has been trained by the bourgeoisie, will not work otherwise."[14]

Lenin's blueprint for relations between the party and the intellectuals, which has remained unchanged for sixty years, made clear to Russian scholars what treatment they could expect in the future in their own country. It is not hard to imagine their dismay at the prospect; but at least one of the "specialists," Professor M. P. Dukelsky of the Voronezh Agricultural Institute, had the courage to protest and, albeit with some reservations, to express his contempt for the regime in an open letter (printed, with Lenin's reply, in the latter's *Collected Works*). This read in part:

I read in *Izvestia* your report on the specialists, and I cannot suppress a cry of indignation. Don't you really understand that not a single honest specialist, if he has retained the least shred of self-respect, can agree to go to work merely for the sake of the animal comforts with which you are offering to provide him? . . . If you don't want to have specialists working merely for the sake of their salaries, if you want new, honest volunteers to join the specialists who are already cooperating with you in some places, not out of fear, but conscientiously, in spite of the fact that they disagree with you on principle on many questions, in spite of the humiliating conditions into which your tactics often place them, in spite of the unprecedented bureaucratic chaos that reigns in many Soviet offices and which sometimes wrecks even the most vital undertakings—if you want all this, then first of all purge your Party and your government offices of the unscrupulous *Mitläufer* [hangers-on, casual fellow travelers], comb out these self-seekers, adventurers, scoundrels and bandits who . . . are either, owing to their despicable natures, grabbing public property or, owing to their stupidity, are cutting at the roots of public life by their absurd, disruptive fussiness. If you want to "use" the specialists, do not buy them, but learn to respect them as men, and not as livestock and machines that you need for a certain time.[15]

This letter represents one of the last attempts of the Russian scientific community to defend its honor and independence. But, as arrests and executions followed thick and fast, those who were still at liberty had no choice but to conform or emigrate. Among the three million Russians who found themselves in exile after the Civil War, several thousand were connected with the arts, literature, or the world of learning. To be sure, Soviet citizens have heard of Bunin or Rachmaninoff, but they have no notion whatever of the activities of hundreds of eminent Russian scientists in foreign countries since the Revolution.[16]

By the early thirties, those of the prewar intelligentsia who had not been shot and had not managed to emigrate existed in a state of terror. Many had hoped that a relaxation of conditions would set in as the Bolsheviks consolidated their power; but years went by, and the regime was more hostile than ever. In 1929 the Academy was purged, and some of its members and staff went to prison. This year also witnessed the beginning of the "Platonov case," when the historian and academician S. F. Platonov and a group of his assistants were arrested and put on trial; Platonov died in exile in 1933. In 1930 another show trial took place, that of the "Industrial Party"—an alleged conspiracy of technicians and saboteurs under Professor Lev Ramzin. No such party actually existed, and after Stalin's death, those who had been thrown into camps and prisons on this charge were rehabilitated. But at the time it served as a convenient way of whipping up public opinion, as the question whether Ramzin and his colleagues should be executed was put to the vote in universities and research institutes.

Out of two hundred staff members at the Karpov Chemical Institute in Moscow, all but two voted for the death penalty. What were the moral principles that prevented the small minority from voting with the rest? Forty-five years later, I. A. Kazarnovsky, one of the two heroes of 1929 and a corresponding member of the Academy, recalled that the vote had been conducted by a professor who was also Deputy Commissar for Justice. "He asked me why I voted against execution, and I replied frankly that the victorious proletariat could afford the luxury of not shooting its defeated enemies."

In the early thirties the CHEKA, now rechristened the OGPU, evolved a new technique of arresting scientists in groups according to their academic specialty. In 1930–1931 a large number of microbiologists were put under lock and key, including Volferts, Golov, Suvorov, Elbert, Gaysky, Nikanorov, Velikanov, and Bychkov. Others arrested were members of the Microbiological Institute at Saratov, lecturers at the Kharkov Medical Institute, and other scholars at Astrakhan, Minsk, Tashkent, and Moscow. Some were put to death soon afterward, others imprisoned in a former monastery in the old town of Suzdal. The monastery was converted into a so-called "Special Institution" in which about nineteen highly qualified academics were set to work developing offensive and defensive bacteriological weapons. The more fortunate of these—Elbert, Gaysky, Suvorov—were eventually released, but those, such as Nikanorov, who could not produce the desired results were shot.[17]

After the microbiologists it was the turn of the agronomists and botanists, who were made scapegoats for the collapse of agriculture due to collectivization. (Mass arrests and executions of geneticists were still to come.) In the early spring of 1933 N. I. Vavilov, director of the All-Union Plant-Breeding Institute, who had just returned from a tour of America lasting several months, wrote to Professor A. A. Sapegin: "The most incredible things have been happening. Twenty of us have fallen by the wayside, including G. A. Levitsky, N. A. Maksimov, V. E. Pisarev, and so on, and nobody knows where it will end."[18] Vavilov's amazement is understandable: the cytologist Levitsky, the physiologist Maksimov, and the plant breeder Pisarev were all men of world reputation, and there was not the least reason to doubt their political loyalty. Yet, before long, Vavilov heard that his correspondent Professor Sapegin, director of the Selection Center at Odessa, had himself been arrested and executed.

And so it went on. Professor V. V. Talanov, the chief Soviet expert on corn, was arrested three times before his death, while V. S. Pustovoyt, a sunflower specialist and a future academician, was arrested at Krasnodar and sent to a prison camp in Kazakhstan. Plant breeders, agronomists, and animal husbandry experts were arrested at Dnepropetrovsk, Rostov-on-Don, Kiev, and throughout the country; rumors were leaked that they had conspired with the kulaks to ruin Soviet agriculture by poisoning cattle, infecting crops, and so on.

The talk of conspiracy was of course pure invention. Even in 1919 there were probably few who believed that academicians and professors were planning a coup d'état, and by the early thirties such accusations were simply ridiculous. But the politicians who worked so steadily over the years to destroy Soviet science did so not only for immediate practical reasons, as in the case of the microbiologists, but also with the long-term purpose of *bringing science under political control*. This aim can be clearly discerned throughout the long history of relations between scientists and the Soviet authorities. Sometimes the latter adopted a direct policy of cajoling the better-known and more "useful" scientists. Thus, in 1921, when Pavlov wrote to Lenin asking for permission to emigrate, as he could not carry on his work in starving Petrograd, the Soviet government issued the often-quoted decree providing him with state aid. But as privileges of this sort cost money and were not effective with the bulk of scientists, it was thought better as a rule to work on them through the CHEKA-OGPU. Here the object was not so much to eliminate the most unruly as to crush, frighten, and demoralize the scientific community as a whole. Even the memory of ordinary decency was to be obliterated: professors and academicians were to be made to kiss the rod, or, in

Orwell's words, to "love Big Brother"; to this end, any and every means was justified.

Professor Victor Evgrafovich Pisarev (1882–1971), a plant-breeding expert, told me that in 1934, when in jail at Saratov, he had been pressed by the authorities to write a denunciation of his close friend and associate Nikolai Vavilov. For some time Pisarev resisted, saying that he knew nothing to Vavilov's detriment; but when threatened that his wife and children and he himself might be killed, he gave in and wrote some rubbish about plots, secret factions, and sabotage in the All-Union Plant-Breeding Institute, of which Vavilov was director. While this was going on, Vavilov himself was doing all he could to get Pisarev released. He wrote to the party Central Committee and to Mikhail Ivanovich Kalinin, "father of the nation," and in the end Pisarev was actually let out of jail. Morally and physically broken, he returned to Leningrad and lost no time in confessing to Vavilov, who forgave him. But the ill repute of an informer clung to Pisarev for the rest of his life, and even after his death, I heard talk in scientific circles of how ignobly he had behaved. This was exactly the kind of effect that the CHEKA aimed at: to set the academic community at odds, to foment quarrels and suspicions, to prevent all unity and sense of solidarity.

The thirties and forties were rich in situations of this kind. Assaulted physically and morally, the old intelligentsia surrendered one position after another. Scarcely anyone was able to retain his post as an academician, a university professor, or the head of an institute, without sacrificing the most elementary decency and descending to one compromise after another.

Here is a typical biography of those days. It starts in 1920, when Ivan Vyacheslavovich Yakushkin, a gifted young professor of plant breeding at the Voronezh Agricultural Institute, took refuge in the Crimea from the advancing Bolsheviks and taught at Taurida University. Later he tried unsuccessfully to escape into Turkey with Wrangel's retreating army. When the Civil War was over, he returned to Voronezh, hoping that his misdeeds had been forgotten, but he was arrested in 1930. Many of his fellows disappeared without a trace, but he survived imprisonment and was given a new lease on political life. As he himself related many years afterward: "In 1931, immediately after my release, I was recruited by the OGPU as a secret agent; I continued in this job until I was discharged from it in November 1952 or 1953."

I came across this disarmingly sincere confession by Professor Yakushkin, member of the All-Union Lenin Academy of Agricultural Sciences, when I was studying the file on Nikolai Vavilov. This Vavilov dossier, bearing the number 1500 and the notation "To be kept in perpetuity," was shown to me in the public prosecutor's office of the USSR during a period of political "thaw" in 1965, and I thus learned not only about the victim's fate but also about the picturesque fortunes of a hired assassin. "As a secret agent of the OGPU," wrote Yakushkin to the prosecutor, Kolesnikov, "I addressed to my superiors denunciations of Vavilov and others." This statement was required of him after Stalin's death in 1953, when the authorities were preparing the rehabilitation of Vavilov, who died in prison. Earlier, while still an OGPU agent, Yakushkin, by his own account, wrote hundreds of denunciations, not only of Vavilov but of Vladimir Komarov (president of the Academy), the physicist Feydor Yoffe, the geochemist Aleksandr Fersman, and many others. In return for these valuable services Yakushkin was dismissed with honor: an informer can rely on getting off scot-free. . . .

Yakushkin, at least, had to be arrested before he was suborned by the OGPU, but in other cases even that was not necessary. The physiologist and academician Aleksei Aleksandrovich Ukhtomsky (1875–1942) was highly respected by his contemporaries as a man and as a scholar. His theory of the dominant figured in all physiology textbooks as one of the major discoveries of the age, and his letters, recently published in the Soviet Union, show him to have been a fascinating personality. But even this worthy man did not escape the process of demoralization. The Ukhtomsky family is a noble and ancient one; it goes back to the twelfth century. This in itself was sufficient to ensure that when the Revolution broke out, their home in the Kostroma province, containing a huge library, was pillaged and burned. Aleksei—together with his brother Andrei, a bishop—was thrown into jail.

Afterward, their paths diverged. Andrei, a fervent Christian, devoted himself to fighting for the rights of the Orthodox Church, while Aleksei indulged his passion for learning. Andrei spent decades in prison and exile, while Aleksei became a professor at Leningrad University and lived in comparative comfort.

But then came 1937, the peak year of Stalin's purge, and—whether at the party's behest or that of the OGPU—at an open meeting of the university, a professor of biology, Nemilov, invited his colleagues to take a look at the record of Academician Aleksei Ukhtomsky. Three sociology students had recently been arrested, and Ukhtomsky had not disowned them or made a public expression of regret and self-criticism. Moreover, he was in secret communication with his brother, a declared enemy of the people, and was sending him material help in his place of exile. Was this not double-dealing?

Ukhtomsky himself was not at the meeting, but when his assistants told him of Nemilov's attack he was filled with terror. At their persuasion he wrote an abject petition to Andrei Aleksandrovich Zhdanov, First Secretary of the Party Committee for the Leningrad District, disavowing his brother and the arrested students. This sad and shameful episode in the history of Russian science might have been forgotten but for the fact that years later, after Ukhtomsky had died of starvation during the siege of Leningrad, one of the three students he had denounced came back from the prison camp. Broken in health, walking on crutches, and desperately poor, he made his way to Leningrad with the intention of paying tribute to his former teacher by writing the story of his life. In the university archives he was given a look at Ukhtomsky's papers. The student, who is now a doctor of science, described what he saw, in a letter to me:

> The draft I read bore witness to the old man's state of terror. He described the three of us as a sly, crafty lot, the very lowest of the low—and he confessed that he had been too old and too simple to realize that we were out-and-out terrorists, as the glorious Cheka had speedily brought to light! I have never in my life felt such grief, or such pity for my teacher as when I held that document in my hands.[19]

World War II was approaching, and the prerevolutionary generation of scholars and scientists had by this time largely died out. Consequently the brunt of Stalin's terror fell on those who had received their scientific training since the Revolution. These younger scientists had no quarrel with the Bolsheviks; many of them were party members, and for the most part they were full of loyalty to the regime as well as to the cause of science. But they too were ground up by the millstone of the state, and thousands of them spent many years in prison, labor camps, and *sharashki* (the colloquial name for prison establishments where scientists worked to order on special projects, as described in Aleksandr Solzhenitsyn's *The First Circle*). Among these were the biophysicist Aleksandr Chizhevsky, the ethnographer Nevsky (who received a posthumous Lenin prize fifteen years after he had died in camp), the epidemiologist Pavel Zdrodovsky, the airplane designer Andrei Tupolev, the physicist Lev Landau, the rocketry expert Aleksei Korolev, the biologists Evgeni Kreps and Aleksandr Bayev, and the physician Aleksandr Parin. Those who eventually returned from captivity made names for themselves in their various disciplines, but how many did not return?

A special group of martyrs consisted of geneticists headed by the academician Nikolai Ivanovich Vavilov. For ten years I collected material for a book on this brilliant biologist, and I can say with pride on his behalf that although he was persecuted to death, his spirit remained unconquered to the last. The founder and first director of the All-Union Academy of Agricultural Sciences, a member of several academies and learned societies throughout the world, a tireless traveler who had visited fifty-two countries in search of cultivated plants, Vavilov endured prison and torture with courage, fortitude, and dignity. During the eleven months from August 1940 to July 1941 he was interrogated four hundred times and nevertheless found time in his cell to write a long work on the history of world agriculture.[20] He was sentenced to death in a court session lasting a few minutes, and was then confined in a death cell for another eight months, where he soothed and cheered his fellow victims and made the best of the time with lectures. This great scientist, who had enriched his country by millions of tons of grain, died of acute exhaustion in Saratov prison hospital on January 26, 1943. In the official record, which I was able to inspect, the cause of death was given as "dystrophy and edematous disease." Other geneticists of those years had an easier death: Levit, Agol, and Karpechenko were shot outright.

But setting aside these dreadful memories, it is a remarkable fact that while the most brilliant and famous scientists were being executed or rotting for years in jails and labor camps, the scientific population of the country was nevertheless steadily increasing.

According to official statistics, the number of scientists in the Soviet Union in 1941 was 98,300, or ten times greater than in 1914,
and by the beginning of 1973 this figure had again multiplied tenfold; at the present time it is claimed to have reached 1,200,000.
There can be few professions whose members increase at this rate
every quarter of a century! We shall have further occasion to discuss the quality of present-day scientists and compare them with
those of sixty or seventy years ago; but first let us try to discover
the conditions and methods which have produced this tremendous
increase in the number of people connected in some way with
scientific research.

The Bolshevik objective of enforcing universal literacy was
probably pursued with more determination and thoroughness than
any of their other aims. Never before in Russia had such vast numbers been sent to school as in the 1920s and 1930s. Of course, like
other political slogans, that of universal education was proclaimed
without any consideration of ways and means. The country was
short of teachers, classrooms, textbooks, and exercise books. I
remember how my mother, teaching at a village school near Vo-
logda, had to cope simultaneously with three classes, housed in a
single large wooden hut. She would assign an essay for the eight-
year-olds, tell the next younger ones to get on with their arithme-
tic, and give a spelling lesson to the youngest class, to which I
belonged. There was one spelling book for every five pupils, so
that we huddled together and looked over one another's shoulders.
I have forgotten how much we succeeded in learning, but at the
end of the day my mother was half-dead with fatigue. Such was
the elementary education, enjoyed by many who are now doctors
and "candidates" of science.*

Teachers' training institutes began to spring up somewhat late
in the day. In true Soviet style a crash program was set up to ab-
sorb as many students as possible and to turn them out, ostensibly
qualified, in record time. The villages were crying out for teachers,
so what could the authorities do but send them second-year undergraduates—they would complete their education somehow, at
evening classes or by correspondence. . . . In this way, ignorant

---

* Two postgraduate science degrees are given in Russia: the first, and
the lower, degree is *kandidat nauk*—"candidate" of science; the second, and
higher, degree is *doktor nauk*—doctor of science. The "candidate" degree
is equivalent to a Ph.D. in the West. —Trans.

teachers trained an ignorant generation—not only of other teachers but of doctors, engineers, and agronomists. The slogan, "We must increase output," applied to specialists in all disciplines, exactly as it did to coal, metal, or galoshes. The results were what might have been expected.

I have before me a book typical of many which appeared before World War II: a compilation celebrating the twentieth anniversary of the Tashkent Medical Institute. After wading through endless tributes to the great Stalin, the party, and Soviet power, the reader comes to a few lines from which it is possible to gather what kind of doctors were really being trained in Uzbekistan during those years. According to the writer of the introductory article, the heads of the Institute took special pride in the fact that "energetic efforts were made to safeguard the firm leading role of the proletarian class of student, . . . and to ensure that students were recruited on a class basis."[21] Scarcely any students were admitted from the families of intellectuals, doctors, teachers, engineers, or white-collar workers of any kind.

Another feature in which the authorities took pride was the emphasis on training students of the local nationality. True, the young Uzbeks did not know Russian and had not received even the rudiments of a secondary education, as had Russian young people in the twenties, but what did that matter? As the author writes, "In the first few years it was actually necessary to lower the entrance qualifications for students of local nationality."[22] Simplified textbooks were produced for them, and they were given simplified lectures, while the central authorities continued to cry: "Faster, faster!" And then, just as with coal, metal, and galoshes, it became the policy to reduce from five to four years the period within which half-baked therapists, surgeons, and pharmacologists were let loose on the Soviet public. "All this," as one of the contributors regretfully writes, "could not fail to lead to a sharp decline in the quality of graduate doctors." But he goes on to say that some of these newly fledged Uzbek house surgeons were promptly appointed temporary lecturers "to assist in the training of further scientific workers and teachers."[23] Within a year or two, the graduates were in charge of hospital wards or departments of the Institute, turning out fresh batches of doctors and scientists in their own image.

The situation was similar throughout the country. At the beginning of 1931, the party's Central Committee instructed the Commissariats for Education and Agriculture to speed up the training of agricultural experts. Vavilov, as director of the All-Union Academy of Agricultural Sciences, considered the matter and produced a plan for the training of 5,000 researchers by the Academy's institutes within the next five years. These young geneticists, plant breeders, soil scientists, and livestock specialists would have to be trained in addition to the normal research work of the institutes and without impeding it. This was no easy matter, but reliance could be placed on the experience of the old generation of scientists, who would spare no effort to train their successors. Vavilov announced the new official requirement to his staff, who promised to do their best to fulfill it. But the authorities were not satisfied. Instead of training 5,000 specialists in five years, the Commissariat for Agriculture demanded that 15,000 or 25,000 should be produced in two years. Their mouthpiece for this suggestion was the deputy director of the Agricultural Academy, Professor Vasily Bursky, who amid general applause delivered this speech:

> "This is the target we propose: to enroll ten thousand agricultural workers and, we hope, about another fifteen thousand scientists. How is this to be achieved? By establishing large scientific-educational state farms, to which we shall appoint the best inventors and managers from all over the country. Thus there will be a concentration of brain power on a single question and we shall be able to push on more quickly, to accelerate scientific progress for decades to come. . . . In the state farms we can combine the efforts of five hundred or a thousand of our best inventors. A group of this kind can advance the progress of agricultural science by several decades. I would never exchange a plant-breeding institute for a scientific-educational state farm, concentrating the experience and energy of hundreds of our best workers."[24]

Forty-five years later, this speech sounds like a ludicrous parody, but at the time it inspired deep conviction, combining as it did all the elements of official ideology and Soviet political wisdom. The idea of crossing a scientific institute with a concentration camp has survived to the present day and still commends itself to the authorities as a convenient way of directing science into the right channels.[25] True, Bursky's idea of mass-producing agricultural experts on state farms was not adopted. Vavilov, then a member of the All-Union Central Executive Committee, was able to put across the simple truth that the best place to train a specialist is in an institute. But many of those who discussed the matter at the highest government and party levels were attracted by the idea of a "class-revolutionary method of instruction." At that time, a postgraduate course for the training of specialists was introduced at the All-Union Plant-Breeding Institute at Leningrad, in which preference was given to those who combined Komsomol (Communist Youth League) and party experience with an unblemished working-class or peasant background. Certainly these included some competent, intelligent people, but the principle of selection and the subsequent handling of the course were highly demoralizing to scientists of the future. The professors and academicians who ran the Institute were required at all costs to turn out a quota of graduates at fixed intervals. The staff in those days were highly qualified, but the ablest professors were powerless against the primeval ignorance of some of their proletarian trainees.

M. I. Khadzhinov, a member of the Agricultural Academy who taught at the Plant-Breeding Institute in the thirties, remembers that in 1938 he was summoned to the local party committee for a brief conference:

"Is Shundenko under your supervision?"

"Yes."

"Why hasn't he presented his thesis yet?"

"Shundenko is illiterate, he doesn't want to study, and he is quite incapable of writing a thesis."

"But it's your business to see that he gets his degree. If you haven't been able to teach him, you must write the thesis yourself!"

There was no arguing with the party's decision. Khadzhinov had to dictate a thesis to his dull-witted pupil, who very soon received his degree and was appointed academic deputy to the director of the Institute.[26]

It is interesting to trace the fortunes of some other half-baked scientists who received higher degrees during this period. I met several of them in the mid-sixties when I was collecting material for my book about Vavilov. By this time they were all doctors of science, members of the Agricultural Academy, holders of high positions. In conversations with them, I came to realize that their brilliant careers were generally based on a foundation of organized treachery. At the time when Lysenko was coming into official favor and the first attacks were being made on the science of genetics, these postgraduate students were a kind of fifth column within the Institute. They soon realized that Lysenko was on the rise and that they would get ahead faster by supporting him than by exerting themselves to study under Vavilov, whose ideas Lysenko opposed. Taking their cue from Lysenko, they lost no opportunity to accuse Vavilov and their other teachers of being illiterate and behind the times. Here is a typical scene from a public meeting at the Institute on May 8, 1937:

> KUPRIANOV (postgraduate student): You are frightened to death of criticism—it touches you to the quick, doesn't it? Why do [Institute professors] Rozanov and Vulf try to put the matter in this way? Because they are in favor of Vavilov's theory—a harmful theory which ought to be torn out by the roots, since the working class has coped with its problems without help from the bourgeoisie, and is now in control and achieving results. . . . The whole country knows about the debate between Vavilov and Lysenko; and Vavilov will have to change his tune, for Comrade Stalin has said that the right way to work is Lysenko's way and not Vavilov's.
>
> DONSKOY (postgraduate student). Lysenko has said it straight out —it's either him or Vavilov. "I may be right or wrong," he says, "but in any case there's no room for both of us." That's putting it fair and square. . . . It is time to realize that we are in a period when the achievements of experimental science must be devoted to the service of the socialist motherland. That is the reason for the sharp conflict and hostility to Vavilov's school.[27]

These words are quoted from the first file I came across in the archives, among dozens of stenographic reports of those years. The director of the Institute, who was shortly to be arrested, had to listen to this kind of admonition by the hour. Referring to the current batch of postgraduate students, he remarked on one occasion: "We can hardly expect any good from *these*. And yet it's a pity—there are some who have brains."

Certainly the group in question contributed nothing to science. Instead, they are associated with some disgraceful and ruinous episodes, including the tragic session of the Agricultural Academy in August 1948. Another remark of Vavilov's provides an apt comment on them: "If people don't possess the genes of decency, there's nothing you can do about it."

The "genes of decency" were eradicated in those days in no uncertain fashion. In the 1930s, the hero of the day, second only to the air force pilot, was the engineer. Youths who could not get into flying school dreamed of being admitted to the polytechnic institute—no matter what faculty, so long as they could share in the national task of construction. In those days, we thought that socialism was just around the corner; all we had to do was to build a sufficient number of factories and power stations to step into a radiant future. Making this happen was the ambition of every graduate from high school; but it was no easy matter to get into the polytechnic institutes, where there were twenty-five applicants for every place. Children of the old intelligentsia stood a much poorer chance than anyone else. There were some, on the other hand, who managed to get in without trouble, and almost without passing exams, and these were the so-called "party thousands."[28]

A chemist, now an old man, who entered the Kiev Polytechnic Institute in 1930 recalls that among those who were welcomed with open arms were the head of a provincial finance department, a party district secretary, a Komsomol secretary, a general's adjutant, and the sister of an army commander. The average age of this new privileged caste was twenty-eight; its educational and scientific background was virtually nil.

In the 1930s, the "party thousands" formed a considerable proportion of the students in all technical colleges; they had more influence than any other group and introduced a spirit that was wholly without precedent. These young men and women, representing the new victorious class, treated their teachers with absolute contempt, interrupting, insulting, and ridiculing them. Lectures and lessons turned into a permanent guerrilla war, in which the "sniveling intellectuals" were harassed by fair means and foul.

An ex-serviceman who was now a student was in the habit, when he presented himself for an exam, of placing a loaded revolver on the table under the professor's nose. In the atmosphere of class hostility and mistrust which prevailed for many years, it happened that a physics lecturer, during an experiment, borrowed a penknife from a student and then asked him casually: "What kind of a Bolshevik are you without a knife?" There could be no question in such circumstances of respect for one's elders, of enthusiasm for learning, or of creative zeal. It was not a matter of acquiring a grasp of knowledge by honest effort, but of extorting it from one's defeated adversary.

When the students reached their final year, there would be a debate on "who is to become a scientist?"; and again the choice would fall to an ex-commissar from the Baltic Fleet, or the female secretary of an important figure. They did not especially covet the honor: teachers and their assistants were poorly paid in those days and did not enjoy great prestige. But the party's wishes were clear, and the Leader insisted that everything depended on cadres, including scientific ones. In 1933 the People's Commissariats, or ministries, began to set up the first specialized research institutes and laboratories. One or the other of the "party thousands" would be summoned by the local committee and told: "Well, Fedya, the time has come . . ." and Fedya would duly embark on a scientific career. Some of those drafted in this way were of course naturally gifted physicists, chemists, biologists, and so on—a wave of this magnitude was bound to wash up a few diamonds as well as stones. But the system of class and party preference, and the emphasis laid on political activity as opposed to intellectual merit, meant that the great majority of those who secured places in laboratories, clinics, experimental centers, and research institutes were not the best, the most honest, or the most capable. Still, the places were filled, and the scientific population began to increase.

Then, after World War II, there was a breakthrough, as Stalin doubled and trebled the salaries of candidates and doctors. Academicians received a special additional stipend of 5,000 rubles a month. The Leader had his own reasons for doing this. His victorious empire was already embellished by the introduction of gleaming shoulder boards for officers and golden crosses on the domes of newly reopened churches, and he wished to add the luster of science and the arts. This of course meant that scientists had to be properly paid.

Another reason for increasing scientific prestige was the prospect of militarizing science and of harnessing it to atomic, hydronuclear, and similar projects. Thus it was that during the hungry forties, officers, priests, and scientists were among the best looked after of the population; and the achievement of a higher degree was a passport to material success. Doctors, engineers, factory chemists, and teachers who had been existing on meager salaries found that there was a magic solution to their difficulties: all they had to do was to compose and present a doctoral thesis. The effect was automatic, even blatantly so: on the very day the degree was conferred, their salaries shot up to a highly respectable level.

As a result, both young and old flocked to obey the call of science. Physicists, chemists, and engineers, even without higher degrees, found that job opportunities had improved beyond recognition. Military laboratories and institutes sprang out of the ground like mushrooms. Anyone who was not a Jew, had no relatives abroad, and was not under interrogation or charged with a crime was welcome to apply for a job in a secret institute, where the pay was better still and the arrangements for defending one's (secret) thesis were less rigorous than elsewhere. It was pocket money!

The Higher Examinations Board[29] functioned with the efficiency of an enormous wind tunnel, turning out up to 400 doctor's degrees at a session. Every year 5,000 new doctors were created; candidates were innumerable. The total number of Soviet scientists doubled in the ten years from 1950 to 1960, and again in the following six years.[30] This could be called the response to a social requirement; it could also be called climbing onto a bandwagon.

Khrushchev, in his turn, was not oblivious to the claims of science. Every time a sputnik was put into orbit, the institutions devoted to rocketry were favored by a cascade of decorations and degrees. The order would come from on high: "In connection with the successful launching . . . , twenty candidates' degrees and ten doctors' degrees are to be awarded to engineers engaged in rocket development." This command was kept secret, and no one thought of criticizing or disputing it. A lucky individual would write a report of a few pages, and the Academic Council would forward a secret endorsement to the Higher Examinations Board, which would duly issue an honorary degree. The total number of doctors and candidates manufactured in this way is not known, but we may suppose it to be quite large.

Thus the growth of scientific cadres was convenient to everyone, from the officials responsible for science to the most junior research officer. A scientific assistant with a wife and family cannot possibly live on 84 or even 105 rubles a month (one ruble is about 90¢). The director of the institute knows this, so do all the doctors on the Academic Council, and so does the assistant's immediate boss, the laboratory head with a candidate's degree. They know it and sympathize, and as long as they can do so without harm to themselves they will help him to get a higher degree and a decent salary. The quality of his thesis may or may not be taken into account—what matters in the last resort is that the man must live. And so the ranks of science are swelled until, as we have seen, they rise above the million mark.

It is said that in the last few years this state of things has aroused concern at a high official level; not because of the devaluation of science, but for quite a different reason; that it becomes expensive. Higher salaries have to be paid not only to "necessary" physicists and electronics experts, but also to a horde of philosophers, historians, and philologists, and the crowd of those who have to be supported in this way grows larger every year. Various ways of keeping the flood within bounds have been considered, such as requiring a minimum number of publications, and not long ago the Higher Examinations Board was reorganized. But the flow cannot be stopped altogether; it is too profitable for all concerned. Apart from its basic function (whatever that may be), for thousands of people, science represents a means by which they can earn an easy, simple, and undisturbed livelihood.

Publicly, at any rate, nobody raises the alarm, and a new slogan, that of the "technological revolution," is used to drum up recruits for the scientific army. This revolution, it appears, is now the key to a happy future, and for it Russia must have more and more scientists. . . . Leaving aside the question of a happy future for the moment, we should examine the relationship between the recruitment of scientists and the advancement of scientific knowledge. This is not so simple as it may seem at first glance. Bigger and bigger sums are being spent on the accumulation of knowledge, but the human and material resources and equipment devoted to this purpose are increasing much faster than the results achieved. In fact, an eminent American sociologist has calculated that to double our knowledge we need to increase the number of scientists a hundredfold. If this is so—and no one has contested this view—the advantage of having a million-strong army in this field would seem to be debatable. At the present rate Russia may have ten million in the year 2000, and how many in 2025? Is not the expenditure too great for any results we can expect?

Once, half in earnest, I was discussing this matter with a group of young candidates in a Moscow research institute. I had been giving a talk on "Why a Scientist Needs a Conscience," and those who came up to me afterward were, as far as I could judge, completely indifferent to ethical problems. Anyway, we stood in the lobby discussing various aspects of the future of science, and I happened to say that I thought we were paying too much for progress. At this point, my young friends suddenly became animated. "Oh, the pay isn't too much," exclaimed a man in a brightly colored tie, whose views were clearly shared by his comrades. "After all, what's 280 a month? And 500 would be better, of course. All of us are making heroic efforts to reach that princely amount!" This aroused general merriment, the point being that a doctor's salary, to which they all aspired, is indeed nearly twice that of a candidate. Seeing that I was dispirited by the conversation, one of the physicists (none of whom was over the age of thirty-five) tried to console me. "Don't be too upset about the purity of Science. She lost her virginity ages ago, and now she's more of a milch cow. That doesn't mean that we aren't interested in our work, but the fact that I love physics is my own affair. Not many people in the Institute care whether I do or not, and it isn't what they pay me for . . ."

CHAPTER 2

"COMRADE DIRECTOR" AND OTHERS

"Not much trouble is taken over their education and intellectual development, as it is assumed that their functions are confined to managing."*

People in the Soviet Union are fond of talking about "Soviet physics," "Soviet chemistry," or "Soviet biology"; this provokes an amused smile on the part of foreigners to whom it has long been an elementary truth that science is one and indivisible. Mendeleyev and Pavlov, Rutherford and Bohr, Wiener and Hans Selye, are all equally part of world science, and whatever does not belong to world science is not science at all. All this is perfectly true, and yet I would maintain that "Soviet science" as a mass social phenomenon is indeed something different from what the West understands by science.

This difference has deep historical roots, going back to the fact that Russian science from its very beginning in the eighteenth century was a political, state-directed affair.[1] This was true of Russian universities and of the Academy itself. By contrast, the Royal Society founded in London a century earlier was called "Royal" because King Charles II and his successors regarded it as an honor to patronize a society of independent investigators. The imperial Russian government, on the other hand, treated the Academy as familiarly as any other of its servants. While professors in European universities taught and researched in an atmosphere of democracy and independence, a Russian professor was never more than a state official of a certain rank. From this point of view, it was logical, when the Revolution came, for the Soviets to over-

* M. E. Saltykov-Shchedrin, *Complete Works*, Vol. 8, p. 241.

throw Tsarist science in the same way as they attacked all other departments of the Tsarist administration.

Having smashed the bourgeois economy, the system of private landownership, and the whole apparatus of Tsarist government including science, the Bolsheviks stopped short at a point which no revolution has so far crossed: their destructive fervor left untouched the cultural stereotype which has characterized Russia throughout the ages. That stereotype, expressed in former times by such institutions as the Army and the Church, is reflected in the present-day organization of the Soviet party and government, the secret police (CHEKA, OGPU, NKVD, KGB), and Soviet science.

American physicist Robert Oppenheimer wrote, with the experience of Western universities in mind, that science was essentially democratic. Soviet science, from its very beginning around 1930, has been the negation of this principle. One of the stereotypic features of Russian culture has always been that everything in the country, including science and the scientist himself, is government property, and that the higher ranks of the community, whether scientific or other, enjoy absolute predominance over the lower. The equality of senior and junior scientists in the face of scientific truth, the right of an assistant to challenge the judgment of his superior, sounds no less ridiculous to Soviet ears than it did in Tsarist times. Soviet science was militarized in spirit and organization long before it was made an appendage of the Soviet military machine.

The typical Soviet institution in this field is the "scientific research institute," independent of a university or a corps of professors. In the West, this type of institution has never been common, and the great majority of valuable discoveries have always been made at universities. The Soviet research institute with its stratified staff—director, deputy director, section head, laboratory head, senior and junior research officers, and laboratory assistants—presupposed from the very outset inequality, subordination, and a barrackslike atmosphere, which makes it natural for the institutes and laboratories to be referred to in the Soviet Union by such military terms as "unit" and "subunit."

The militarized structure required that workers at every level should have precise functions assigned to them and should conform strictly to their place in the hierarchy. The Bolsheviks did not want an association of thinking human beings who might be tempted to disobey instructions from above. At the outset, it is true, many academic institutes were headed by genuine scholars; but it soon turned out that academicians and men of ideas—men like the physicists Vladimir and Ipatieff and Feydor Yoffe, the chemist Aleksei Chichibabin, the geochemist Vladimir Vernadsky, the geneticist Nikolai Vavilov, the microbiologist Vladimir Omelyansky, and the mathematician Ivan Krylov—were not the type that the authorities required, as they could not grasp that a director's function is not to have ideas but to transmit directives. Not surprisingly, the experts who were not content with this role soon disappeared from the scene. For example, the agrochemist Dmitri Pryanishnikov insisted on the need to build factories to produce nitrogen fertilizers, without which the soil would become exhausted and harvests would dwindle year by year. But the authorities, for reasons of their own, did not want to do this. Naturally they knew best, and in due course they found a compliant expert (Vasily Vilyams) to declare that no fertilizers were needed, that the Russian land was big and rich enough to bear all manner of crops without them. As a result, Pryanishnikov was thrust aside and condemned as an irresponsible babbler, while Vilyams became director of an institute and was extolled as the champion of true progress. As for the fact that the soil did become exhausted and Russia had to buy grain from America—why, that happened under a different lot of officials, so the previous team could not be blamed . . . anyhow, it became necessary to do without academic celebrities, for moral as well as scientific reasons.

In 1911 Lev Casso, the Tsarist Minister of Education, called in the police to quell a student riot at Moscow University. The professors were angered at this, and 124 of them handed in their resignations and left the building. Eighteen subsequently returned and asked to be reinstated, but the great majority remained on strike rather than defer to the ruffian minister. The Russian professor of those times knew his own worth, and was a personality even when in uniform; he knew that society as a whole was on his side and that he was regarded as something special—even unique. Those who took charge of Soviet science learned from such incidents and made it their first task to ensure that the new science was impersonal and independent of the mental and moral qualities and abilities of any single professor.

To this end they first took in hand the Academy of Sciences. In 1927 the Council of People's Commissars doubled the number of academic posts, and at the same time introduced a new voting system. Thanks to these "benefits," they were able in the following year to pack the Academy with a large group of party nominees ("a whole herd of Trojan horses," as one who remembered those times described it to me). The authorities went all out to give the Academy a party complexion and thus create a cadre of amenable institute directors. G. M. Krzhizhanovsky, who had been a friend of Lenin's, was first promoted to the rank of an academic philosopher and then transferred to "stiffen" the physico-mathematical section. Ivan Luppol, a high party functionary of limited education, was also pushed into the Academy (he was to die of starvation years afterward in a labor camp).

The climax came when the academicians were asked to admit into their ranks I. I. Skvortsov-Stepanov, who had been a party member since 1896 but had never completed high school. Apparently they refused in no uncertain terms, whereupon the authorities tried a diversion by appealing for aid to the famous physiologist Ivan Pavlov, who, by this time, had accommodated himself to the Soviet regime, which had presented him with laboratories and generous grants for experiments. Accordingly, at the next general meeting of the Academy, Pavlov made a speech to his colleagues reminding them that the Emperor Caligula had been determined to make his horse a consul despite the objections of the Roman Senate. "But that was a horse, gentlemen," Pavlov exclaimed, "and if you look at Ivan Ivanovich Skvortsov-Stepanov, you can see he's a perfectly decent chap—so why should we make difficulties?" This "recommendation" persuaded the academicians to swallow their prejudices, and Skvortsov-Stepanov was unanimously elected.[2]

The filling of scientific posts with party members, and the gradual disappearance of genuine personalities, naturally made science more amenable to political control. None of the new appointees demanded to be sent on expeditions to distant lands, as Vavilov had done, or, like Pryanishnikov, insisted on the erection of factories for the manufacture of nitrogen fertilizers. Each academician and professor knew his place and valued it too much to disturb the authorities over trifles.

It was, however, many years before all the great scientific

figures died out or were eliminated and replaced by docile nonentities.[3] Even today there are about a dozen genuine scientists occupying directors' posts—men who formerly inspired others with ideas, but who by now have completely adapted to the new system. The change in their functions and their reduction to the status of officials is wholly acceptable to these survivors of former times: they have received Lenin and Stalin prizes and the Golden Star of a Hero of Socialist Labor, and they now live cozily in country villas and are sent on foreign trips by the government.

But the divorce from true science has turned them into eunuchs incapable of creative effort. Among them may be mentioned the biologist Vladimir Engelgardt, the chemists Nikolai Semenov and Andrei Nesmeyanov, the physicist Dmitri Skobeltsyn, the mathematician Ivan Vinogradov, the geneticist Nikolai Dubinin, and the mechanical engineer Ivan Artobolevsky.

The position of these ci-devant scientists in their institutes and laboratories is something like that of a Western constitutional monarch: they reign, but they have scarcely any responsibility for events, and few obligations to science or to their subordinates. A sociologist of my acquaintance was able to read the minutes of academic councils and conferences held by Vladimir Engelgardt, the director of the Institute of Molecular Biology of the USSR Academy of Sciences. At one of these, Engelgardt, who had just come back from an enjoyable trip abroad, said to his laboratory heads, "Tell me something of what you're doing—it's a long time since I knew, and I don't even understand the work you're on at present." Engelgardt is one of the directors who are most firmly anchored to their jobs; he is highly regarded in party circles.

The position of director of a scientific research institute is one of the most highly coveted in Russia and is conferred only as a reward for special services to the party and government. The lucky man may not necessarily be a party member, but he must show absolute obedience and cooperation. If he deviates in the slightest degree—even unintentionally—from the wishes of the district or municipal party committee, let alone the Central Committee of the Soviet Communist Party, he is punished with immediate dismissal, whatever his past services to science or to his country. From time to time the party authorities devise special tests of the loyalty of senior scientists, either individually or as a group. For instance, it may be announced that the head of the department of the Central

life that he had himself written and edited no fewer than ninety books. But times have changed, in this as in every other way.

The custom of directors tacking their names on to the work of others has produced some intellectual giants and prodigies of activity. How many learned publications can a researcher turn out in the course of his life? According to B. N. Volgin, a candidate of technical science, an investigation carried out in thirty-one institutes in Leningrad showed that the average researcher produced one paper a year; others might publish up to ten, but for every one of these there were dozens who published none at all.[5] If this is taken as applying to all academics, including those of advanced age, it would seem that in thirty or forty years a scientist might publish a maximum of 300 to 400 works. But at the present time it has been shown that most directors of institutes have 500 to 600 or even more to their credit, including a large number of monographs. The youngest academician, Yu. A. Ovchinnikov, director of the Institute of Chemical Compounds of the Academy of Sciences, has published 300 works in fifteen years, including several books. But the greatest marvel of all is Andrei Nikolayevich Nesmeyanov, who in forty years has put his name to no fewer than 1,200 works, at the truly amazing rate of an article or monograph every twelve days. That is what it really means to be a director!

In our day the head of a research institute is poised to achieve great things in many spheres. For example, if he wishes to become famous by inventing something—a machine, a new drug, or a means of increasing egg production—he is free to do so and will have no trouble in getting an inventor's certificate and even an international patent, not to mention a Lenin prize, without so much as lifting a finger. I could give dozens of instances; the explanation is simply that under the etiquette of Soviet science, whenever an invention of any importance takes place in a research institute, the director is named among its authors. Otherwise, even the most brilliant invention will not be recognized, since neither the party district committee nor the ministry to which the institute is subject will back it if it does not carry the director's name.

Once, it is true, I came across the tragicomic story of a director who ruined himself as an inventor—but through no fault of the authorities. Not long ago, in a Moscow academic institute, a commission of the Committee on Inventions and Discoveries inspected two versions of a newly devised machine. The committee representative thought well of the first version, but the director, an academician, criticized it roundly as an imperfect, ill-designed, and worthless model; the second version, on the other hand, was excellent.

"Excuse me," said the committee representative in amazement. "The first model may not have been fully worked out, but you yourself are listed as one of its authors, so we thought . . ." There was a ghastly pause, reminiscent of the final tableau in Gogol's *The Inspector General*. As everyone present realized, the director had so often and so shamelessly taken the credit for others' inventions that on this occasion he had mixed them up and claimed the wrong one.[6]

This story shows that even directors have their problems, but on the whole these quasi-scientists lead a very comfortable life, as long as they are not dismissed. That may happen at any time but, especially in Academy institutes, they have no reason to fear removal on grounds of age. I remember that A. N. Bakh, when eighty-nine years old and in his dotage, could not even recognize his first pupil and deputy, Academician Aleksandr Oparin, but he nevertheless remained in his post until death. Lack of ability or creativity is not a ground of dismissal either, and immoral behavior in public matters is not a very damaging offense.

When I was studying the archives for my biography of Vavilov I found several denunciations of him by members of his staff in the 1930s. Thirty years later one of them, S. Sidorov, was still working in the Plant-Breeding Institute and had become its deputy director. In 1967 I gave an address to the Institute staff and read out the denunciation to them. Sidorov was forced to leave the Institute, but the Leningrad District Committee of the CPSU, which was extremely annoyed by my action, immediately made him deputy director of another institute.

As for a director's private morality, the party bosses will not hold that against him unless it comes to corrupting minors. There is, however, one extremely dangerous area in which a director can easily ruin his own career. In no circumstances must he diverge from the party line or deviate by a hair's breadth from what the party considers to be his duty.

In the summer of 1975, in Leningrad, I witnessed the downfall of the well-known academician and physiologist Yevgeny Mikhailovich Kreps. Like many Soviet scientists, Kreps, who was born in

1899, had a checkered career. During Stalin's regime, he spent some years in the camps at Magadan (northeastern Siberia), where he contracted a severe spinal disease. Later he returned to Moscow and was widely acclaimed. In 1961 he was summoned home urgently from an official trip abroad to take the place of L. A. Orbel, who had just died, as head of the Institute of Evolutionary Physiology and Biochemistry in Leningrad. Five years later, Kreps was elected a full member of the Academy of Sciences, and in 1967 he became Academic Secretary of the Department of Physiology. On the two hundred fiftieth anniversary of the Academy, he received the country's highest decoration, the Golden Star of a Hero of Socialist Labor. Kreps was extremely adaptable and gave the authorities no grounds for complaint; as a scientist, he was respected by his colleagues and was even to some extent known in foreign countries. But all this collapsed in a moment when he fell foul of Danilov, the party secretary in the Physiological Institute. Danilov was neither an academician nor distinguished in any other way, but he carried far more weight in the Institute than its director. Kreps was called before the Vyborg district committee to account for himself, and its first secretary, V. M. Nikiforov, asked him bluntly if he did not think he had better retire. Kreps, who was seventy-five, replied that he still had some scientific work to finish. Not long afterward Nikiforov again summoned Kreps and put the same question to him; at the same time he spread a story through party channels that Kreps was refusing to take advice. The Central Committee took up the cry, and Kreps was promptly dismissed by Mstislav Keldysh, the president of the Academy. (The Academy statute provided, it is true, for reelection, but this had no bearing on the present case.)

What misdeed had poor Kreps committed? To the uninitiated it might seem a matter of no consequence whatever: he had shown reluctance to conclude what is called a "patronage agreement" with a state farm outside Leningrad. The essence of such agreements, as opposed to pious talk about the solidarity of town and country workers, is that the state farms haven't enough hands to gather in the harvest, and every year scientists are expected to spend a few weeks on the land picking potatoes. No one in Russia is surprised by this arrangement as such: millions of townsfolk annually quit their offices, factories, and institutes, replacing peasants who have themselves migrated to the city. But Kreps, the Hero of Socialist

Labor, thought himself entitled to override the system by which
doctors and candidates of science are sent to dig potatoes. Basically he was right, but that did not help him. The party district
committee regarded patronage agreements as a key factor in its
control of science, and no institute must be allowed to disregard
them. Kreps showed recalcitrance and was overthrown; his successor, very wisely, made Danilov his own chief deputy.

Kreps's story is an elegant demonstration of crime and punishment. Not long before, however, the same Vyborg district committee had displayed its power over another scientist, less openly but
no less decisively. In 1964 the Academy elected as a corresponding member Mikhail Ivanovich Budyko, aged forty-four, a
gifted meteorologist and expert on the geophysics of the atmosphere and hydrosphere, who soon became director of the central
geophysical observatory in Leningrad. Budyko's star continued to
rise: he was elected to the committee of the department of oceanography, physics, and geography, and was nominated for full
membership of the Academy.

Then came the bombshell. At a general assembly at which the
election was about to take place, Keldysh, the president, read out
a letter from Nikolai Romanov, first secretary of the Leningrad
district committee of the party, informing the Presidium of the
Academy in angry terms that Budyko in his capacity as a director
had ignored the advice of the party organization within the Institute and had come into conflict with its secretary. Accordingly,
Romanov did not recommend that Budyko be elected to full membership of the Academy. True, the latter body was not precisely
subordinate to the party's district committee, but a majority of
those present preferred to conform to the party's wishes. Budyko
was not elected, and before long he was removed from his post at
the Observatory. His crime lay in the fact that, contrary to a secret
party order, he had appointed some Jews to his staff. Such "independence" could not be tolerated; he was dismissed and succeeded
by the party secretary who had reported him to the district committee.

These two episodes have a common feature. Since the reasons
for the dismissal of a director are generally shrouded in secrecy,
the staff of the Institute of Evolutionary Physiology were for a
long time uncertain whether the district committee was responsible
for the removal of their chief. It was hard to believe that petty

party officials could deal so casually with scientists of world reputation. Various conjectures were bandied about in the corridors of
the Institute: perhaps Kreps was partly to blame, perhaps he had
lost his temper with the district committee or offended the Academy higher-ups in some way. But one day, at a routine party
meeting in the Institute, a young instructor from the district committee stood up and declared that it was indeed that body which
had ousted two directors of Academy institutes. The operation
had been carried out deliberately and was a credit to the party
workers concerned. The scientists would do well to take note and
to bear in mind who was really in charge of their affairs.

However, relations between science and the authorities are not
quite so simple as those between a horse and its rider; there are
nuances.

A man who lives all his life under coercion cannot remain in a
permanent state of tension. He cannot regard himself perpetually
as a victim or a freedom fighter; he strives to achieve a more bal-
anced and psychologically comfortable relationship with the outside world. He starts to look for explanations and even justifica-
tions of oppression, and endeavors to convince himself and others
that it can be endured somehow. As an Estonian sociologist wrote
to me after his laboratory at Tartu had been disbanded and destroyed: "We have been through worse things—the siege of Leningrad, for instance."

Thanks to this kind of modus vivendi, people accommodate
themselves to the more disgusting aspects of their existence and
even to some extent cooperate with their tormentors, not necessarily out of cynicism but with the best of motives—for the sake of
science itself. N. N. Semenov, a chemist and winner of the Nobel
prize, defended this attitude to one of his closest associates, with
particular reference to the appointment of party workers to academic jobs. Nowadays, he pointed out, it was impossible for a scientist to do the work of directing an institute or a section of the
Academy—the administrative burden was so great that he would
have no time or aptitude for creative work. Hence there was no
harm in giving such jobs to party comrades or other officials who
had the right administrative skills. Of course that meant they
would have to be given academic degrees and titles, but how much
valuable time it would save the genuine scientists!

Semenov practiced what he preached. For example, as vice-president of the Academy he voted to confer correspondent membership on a forty-five-year-old economist, Vladimir Alekseyevich Vinogradov. He did so although he was perfectly well aware that Vinogradov knew nothing about economics: he was a senior officer of the KGB who had for several years been deputy director, and then director, of the Academy's foreign department, the functions of which were well known. Semenov was not perturbed by this—he believed he was acting rightly as a matter of principle. On another occasion, K., a chemist and a corresponding member of the Academy, asked Semenov during a walk in the grounds of the sanatorium at Uzkoye why he was supporting the claims of a certain Kh. to high office in the Academy, although everyone knew the man had no scientific attainments whatever. Semenov replied, "Yes, he's a nonentity in science, but he's high up in State Security, and it's important for us to have someone there."

Semenov is neither a cynic nor a careerist—he is a scientist of world repute and has received the highest honors in the Soviet Union. But he is not the only scholar to be convinced that the Academy leadership cannot subsist without the aid and participation of high-ranking KGB members and party bosses.

In my early days as a reporter, I once dropped in at the head office of the Soviet Red Cross. I was received by a dark-haired, lively-looking man who began to tell me interesting tales of his work as a medical aide in Ethiopia, Iran, and other distant lands. He had apparently spent many years organizing Soviet hospitals in Asia and Africa, bringing disinterested aid to the peoples of developing countries. Absorbed by the details of his unusual career, I said I would write about him in one of the magazines that the Soviet Union produces for foreign countries. But he replied with a touch of irony, "I shouldn't if I were you. They know too much about me already."

I understood this strange remark several years afterward, when I learned that the official in question, O. Baroyan, was a colonel in the KGB and a specialist in espionage and subversion, whose hospital work had served as a cover for activities that had nothing to do with medicine. He had finally been excluded from the World Health Organization and had decided to find himself a quieter post. In this he was helped by no less a person than Lev Aleksandrovich Zilber (1894–1966) of the Academy of Medical Sciences, a brilliant virologist and pioneer of the virus theory of cancer, who saw to it that the KGB man received a doctor's degree and was made a corresponding member of the Medical Academy. When Zilber, a man of fiery temperament and a perpetual fount of new scientific ideas, was asked why he had helped "that scum" to make his way in the academic world, he laughed and replied, "Even scum has its uses." Zilber, a Jew who had been in prison three times under Stalin, was afraid that if he did not keep on the right side of the authorities, they might again prevent him from carrying on his work, and for a time Baroyan served as a kind of lightning rod. But in due course Zilber died, while Baroyan stayed on. Today he is a full member of the Academy and director of an important research institute, that of epidemiology, microbiology, and immunology, named for N. F. Gamaleya (1859–1949).

Having become a big boss, the former spy and saboteur Baroyan is now actively engaged in sabotaging science. The system that prevails in his Institute is one of unmitigated terror. On several occasions members of its staff who are doctors or candidates of science asked me to write in the papers about Baroyan's misdeeds. Ruling the roost in his double capacity as an academician and a member of the KGB, he bullies his subordinates on academic councils, dismisses anyone he doesn't care for, and treats with special ruthlessness the laboratories carrying on Zilber's work. Of one of his victims, a talented doctor named T. Kryukova, he declared publicly in 1974: "I'll take her back into the Institute if she applies to me this very day and kneels in my office. If she waits till tomorrow, she'll have to crawl to my office all the way from the lobby." Much might be said of an academician of this kind, but perhaps even more blame attaches to the man who helped Baroyan to his current position.

At the present time, many scientists are still appointing successors of this kind; some out of cynicism, others out of negligence or moral cowardice. One of the chief offenders is the economist Abel Gezevich Aganbegyan of the Institute of Economics and Industrial Organization of the Academy of Sciences (the Science City at Novosibirsk). Aganbegyan has made it as easy as possible for local party leaders to present dissertations in his Institute, including those that qualify for a doctor's degree. Not surprisingly, he enjoys the warm support of members of the city and district party committees.

Whereas Semenov acted in accordance with a theory, while

Zilber promoted Varoyan as a matter of calculation, Aganbegyan with his sociological background has perceived an important feature in the psychology of party bosses: they want to consolidate their privileges and increase their security. Having created a kind of *dolce vita* for rectors, prorectors,* and directors, one day the party officials noticed that although scholars' privileges were less than their own, they seemed to be more stable. The life of an institute director or university head was much safer and less disturbed than that of an *apparatchik* (bureaucrat) who might at any moment be hurled from his position at the whim of his superiors. A doctor's degree and, still more, the rank of academician were a far better protection against such storms than any job in a ministry or district committee—in fact, they were the best form of insurance available within the Soviet system.

Having grasped this fact, the bureaucrats started invading the scientific terrain. Ministers led the way. B. V. Petrovsky, a surgeon and Minister of Health of the USSR, got himself elected to the Academy of Sciences, although according to its constitution it does not admit members of the medical profession. To be on the safe side, he also founded a research institute in which he holds the rank of director, though he is hardly ever seen there. The Minister of Higher Education, V. P. Yelyutin, also made an attempt to join the Academy, but in vain. Before the vote, an academician asked about the minister's scientific qualifications. To the surprise of those present, including metallurgists, it turned out that he was the head of a laboratory in the Steel Institute. "But what exactly has he done personally?" the questioner continued. The secretary of the department was at a loss for an answer, and laughter broke out in the hall. So far Yelyutin's ambition is unrealized; but this is an exceptional case, and it has not deterred hordes of officials from attempting to storm the heights of the scientific Olympus.[7]

Deputy ministers are no less anxious to plan for a rainy day. For example, every deputy minister of health manages, in that capacity, to secure a doctor's degree at the very least, and also to select an institute to which he can migrate as director whenever his political fortune turns sour. For that purpose, he also has to get elected to the Academy of Medical Sciences. In this he is usually successful, so that at the present time our deputy ministers all

* "Rector": head of a university; "prorector": deputy head. –Trans.

rejoice in academic titles and have earmarked institutes and laboratories of which they will, in the future, be the directors. Not long ago, this ploy was the subject of a special debate in the Council of Ministers of the USSR. Toward the end of 1975, just before the Twenty-fifth Party Congress, the Central Committee received tens of thousands of letters from workers with complaints of all kinds, a good half of which were about the deplorable condition of the state health service. Petrovsky, the Minister of Health, drew up a report for the Council of Ministers, but this did not satisfy his superior, a member of the Politburo named Kirill Mazurov. Expatiating on the reasons for the collapse of the health service, Mazurov mentioned among other things the strange predilection of the minister and his deputies for acquiring academic titles and scientific posts. Mazurov reportedly shouted to the doctor-bureaucrats that they were not put in office to carve out a scientific career for themselves but to give the country a good health service. The officials of course swore to mend their ways, but none of them abandoned the institutes they had laid claim to.

In recent years, the "academic" craze has spread to other disciplines. Senior bureaucrats in the field of chemistry and geology, railroad experts and agronomists, have no trouble in presenting doctors' theses, as they have under their authority branch institutes in which tens or hundreds of the staff, of their own free will or otherwise, are engaged in writing theses for their superiors. Officials of Gosplan (the state planning agency) and the Central Statistical Office are specially attracted to a scientific career, and the task of composing a thesis is made easier for them by the profusion of facts and figures that makes its way onto their desks in the ordinary course of official work.

The supreme party bosses, who are practical men, have actually regularized the system whereby degrees are conferred almost automatically on up-and-coming officials. Every graduate of the Academy of Social Sciences receives the degree of a candidate of philosophy. True, he must then see for himself about obtaining a doctor's degree: these are not yet standard issue.

The craze for party and state officials to obtain higher degrees did not grow up over night: its first beginnings can be traced from the early fifties. In the mid-seventies, however, it reached epic proportions. "I am glad to be leaving the world of science at a time when bureaucrats are invading it"; these were the words, shortly before his death in 1974, of Nikolai Vasilyevich Lazarev, a pharmacologist who is the hero of several of my books. After 1974 I made extensive trips throughout the country and was able to verify that the "invasion" was in full swing. At the Agricultural Institute in Krasnodar, which I had visited several times in the past twenty years, I found the staff in a state of depression, as the party's regional committee had appointed a new director. This man, I. V. Kalashnikov, had been in charge of the committee's agricultural department eight years earlier; he had displayed neither knowledge nor ability, but had gotten his subordinates to collect and write up the material for a dissertation. Thanks to their work, he became a candidate of economic science and now, as director, was in charge of the work of dozens of experts. In 1975, also at Krasnodar, a new director of the Institute of Civil (Agricultural) Aviation was appointed, a former chairman of the regional committee, who had similarly become a candidate of economic science thanks to the work of his grateful staff. It is true that this man did not become a scholar entirely by his own volition: the story went that his drunkenness and debauchery were such that the regional committee decided to shuttle him off into a scientific career, where his habits would be less noticed.

It is a long way from the fertile Kuban meadows to the slopes of the Urals, but the situation in both places is the same. For seven years the artists and intellectuals of Sverdlovsk, a city of a million people in the Urals, were plagued and bullied by M. S. Sergeyev, the third secretary of the party's regional committee; Sergeyev was in charge of all cultural affairs including theaters, publishing, newspapers, etc. His boorishness was outstanding even among his party colleagues; he effectively throttled all cultural activity and was cordially hated by actors, writers, journalists, and painters alike. Then one day came the joyful news that this apostle of culture was being transferred—into a scientific career! It had been arranged in the usual smooth, inevitable fashion: first an institute had been selected—that of economics, under the Ural Scientific Center of the Academy of Sciences; then an economic thesis had been produced to order, and by 1975 Sergeyev was vice-president of the Center. Since its president (S. V. Vonsovsky) is an academician, it is only proper that Sergeyev should at least start off as a corresponding member of the Academy, and we may safely assume that he, too, will achieve this rank.

Science affords excellent jobs for the big shots, and for those less senior there are also quite pleasant rewards to be had. Myznikov, head of the propaganda division of the Sverdlovsk district committee of the party, lost his job on account of debauchery, but he did not remain unemployed: he is now a candidate of historical science and a lecturer in the party school. When he meets former colleagues he boasts of his good fortune: life is easier and gayer, there's scarcely any work, and the pay is as good as in the district committee. And he is telling the truth. Panfilov, editor of the Sverdlovsk evening paper but better known in town as the "sexual democrat," realized that his drinking and womanizing would get him into trouble someday and decided on a law career. This indeed proved to be his salvation: he has a candidate's degree and is a teacher at the Sverdlovsk Law Institute, instructing young people in the fundamentals of legality and jurisprudence.

The list could be prolonged with examples from Voronezh, Vladivostok, Novosibirsk, and so on—but perhaps the point is adequately made. Officials have decided to become scientists—there is no way of stopping or dissuading them—and the future of science is thus doubly in their hands. But what effect does this have on genuine members of the scientific community?

Here I would like to quote two Moscow scientists: G. I. Abelev, doctor of biology at the N. F. Gamaleya Institute of Epidemiology and Microbiology under the Academy of Medical Sciences of the USSR, and his wife, E. A. Abeleva, a candidate of science who works in the Institute of Biological Development of the Academy of Sciences. In an unpublished article titled "Ethics as an Element in the Organization of Science," these authors wrote:

> If a scientist has good reason to know that he occupies his post by right and deserves the recognition of his colleagues, this gives him a sense of stability, independence, inner freedom, and equilibrium: that is to say, it creates conditions in which it is easy to hear the voice of conscience. A man in this position values honor and conscience; he feels himself to be a part of science, an active agent and instrument of it. . . . The function of a scientist in such circumstances is simply to be himself, to follow his own interest and con-

science, to seek after justice on the basis of his own inner feeling and experience. In such cases it can be assumed that professional ethics are safeguarded in every field.

Things are quite different, however, if a man occupies a place for which he is unfitted. . . . In this case, he has no inner confidence in himself or his own judgment. He lacks independence both inwardly and outwardly, as he owes his position to someone else. He has to seem other than he is, to play a part instead of living, to obtain recognition by fair means or foul and to select staff who depend on his favor. A man in the wrong place inevitably breeds others like himself, as they are the only kind on whom he can rely. . . . Such a man is an enemy of scientific ethics. For ethics require a natural structure of science, a natural hierarchy among its members based on scientific authority, or rather authority according to the function exercised by a scientist or an administrator. A man in the wrong place is far more harmful than a mere vacancy—he is a force of desolation, creating a zone of death around himself. The presence of such people means at times that the ordinary norms of scientific ethics become unattainable ideals, the pursuit of which requires true moral courage, risk-taking, and energy.

It will be clear from this chapter what kind of person the Abelevs had in mind when they spoke of "men in the wrong place."

Let us consider a further question: among the crowd of intruders, would there be any chance in the Russian system for an "outsider" of genius? For example, could the great Albert Einstein have become a candidate of science in the Soviet Union?

In 1905 Einstein, who was then an unknown employee of the Swiss Patent Office at Bern, embodied the conclusions of some articles in a twenty-one-page dissertation titled "The New Determination of Molecular Dimensions." This was approved by two professors of Zurich University, and in the same year Einstein received his doctorate. Let us now try to imagine what would have happened to him in Russia seventy years later.

In the first place, Einstein would have had to pass the three preliminary tests for a candidate's degree, the so-called "minimum," including a foreign language and philosophy. Einstein's English was bad even after twenty years in the United States. As to the philosophy test, he would have found this practically insurmountable: even in his later years, when his genius was recognized throughout the world, he was decried in the Soviet Union as an out-and-out idealist. What is more, he was a religious man, and there is no place for such in Soviet science.

But let us suppose he nevertheless achieved the "minimum." The Patent Office would of course not have an Academic Council of its own, and would have sent his dissertation to the university, accompanied by the necessary certificate concerning his loyalty to the party, sense of civic consciousness, standing among his fellow workers, etc. In all probability, Einstein, a reserved character absorbed in his scientific work, would not have fared very well under all this. Moreover, it is known that Galler, the head of the Patent Office, did not take kindly to Einstein's revolutionary ideas. In Soviet conditions this would have made it quite certain that Einstein was not allowed to present his dissertation.

Supposing, however, that Einstein was allowed to present his thesis and to defend it. Its theme was an explanation of Brownian motion, based on the assumed existence of atoms and molecules. But Ernst Mach and Wilhelm Ostwald, the leading scientists of the day, denied the physical reality of atoms and molecules. In 1905 this fact made no difference to Einstein's academic fortunes, but in Russia in 1975 it would have been a black mark against him, and the abstract of his thesis, circulated according to regulations, would have had a poor reception. As for the "practical application" of his discovery, the less said the better. Consequently, in accordance with current practice and the standing instructions of the Higher Examinations Board, it may be assumed that the university's Academic Council would have rejected Einstein's thesis.

Let us nevertheless suppose that the Academic Council approved the thesis and forwarded it to the Higher Examinations Board. It is most unlikely that the Board would have approved it within a shorter time than three months, as is usual with dissertations of minor importance for the candidate's degree. In all probability, it would have been referred, in view of its unorthodox character, to some behind-the-scenes critic of the Mach-Ostwald persuasion, and one or two more anonymous critics would have had to pronounce on the question of its practical application, if any. All this would have delayed the Board's decision by two or three years. Finally, the Board would of course be bound to notice that Einstein was a Jewish name, and this would have prejudiced his chances still further.

In short, it is obvious that Albert Einstein would have stood absolutely no chance of becoming a candidate of mathematical and physical science; he could not possibly have come up to the standard of a Soviet scientist of the 1970s. Moreover, if he had lived in Soviet conditions and known the tribulations in store for him, he would probably not have tried to present his thesis in the first place. He was, after all, a busy and serious-minded man, but not an ambitious one.

CHAPTER 3

FORCED LABOR OR TAX?

"But the townsfolk were equal to the situation, and very cunningly opposed the forces of inaction to those of action."*

Those whose business it is to look after Soviet science may be easy in their minds: order prevails in the 5,000 scientific establishments of the USSR, and there is no reason to fear an invasion of Einsteins. The 15,000 to 20,000 rectors, prorectors, directors, and department heads, not to mention degree-holding party secretaries on special assignment, keep a firm grip on the 1,000,000 scientists under their control.

Say what you will, 1,000,000 is an impressive number. True, the picture looks different from different angles: the authorities take one view of it, while each of the 1,000,000 individuals takes another. The rulers of old Russia took a personal pride in the Imperial Academy of Sciences and the imperial universities, as a proof that Russia was as good as Europe. Science was a jewel in the imperial crown, a bauble whose purpose was to adorn and give importance to the state. Russia's present rulers, too, have their own ideas as to what science is for. In every speech they are careful to insist that it is not an end in itself, but is part of the state and is there to serve official purposes. It serves, and receives its wages. Government and party leaders who take the floor at academic ceremonies refer to science in the tone of a benign employer commending a satisfactory workman. For instance:

The Communist party and the Soviet state . . . highly value the work of eminent scientists and, as you know, they show their appreciation in a worthy manner. . . . But in future, comrades, we must work harder still, with more determination and even greater

* M. E. Saltykov-Shchedrin, *Complete Works*, Vol. 8, p. 338.

efficiency. More than a million workers are engaged in different branches of Soviet science. This is a great force, and it is of the utmost importance to use it properly. Altogether, comrades, I would say this: the higher our party values the work of scientists and their part in building Communism, the more it expects and the greater demands it makes on them.[1]

And so on, and so on. But the note of thrift and economy which can be heard in speeches like this does not apply to practical life. Russians live in a political state under rulers who are guided, not by economic or even ideological considerations, but by those of day-to-day politics and immediate pragmatic advantage. Relations between the authorities and scientists are also governed by politics. In the view of many well-known economists and philosophers of science whom I have consulted, the great majority of the country's army of scientists—1,169,000 in 1973, and now 1,200,000—are not needed at all. Russia has overdone things in this field as in many other kinds of specialized activity: the USSR has two and a half times as many engineers as the United States, and an overproduction of chemists, physicists, teachers of history, and so on. The structure of Soviet science is reminiscent of the ancient Orthodox cathedrals that may be seen in Kiev or Vladimir, with their astoundingly thick walls, massive ceilings, and buttresses. Today's architects know that churches do not need to be built like fortresses, but the builders of the past knew less than we do about the strength of materials and preferred to have a safety margin. The rulers of present-day Russia have erected an imposing structure of scientific organization on the same medieval principle, partly with an eye to propaganda—"You see, every fourth scientist in the world is ours!"—but also, and chiefly, as a margin "to be on the safe side."

Certainly the margin is a wide one. When the Japanese wanted to build a monorail connection between Tokyo and its airport, they invited eight or nine German engineers to advise them, and the job was done. In Russia, for the same purpose, a whole special laboratory has been created. Russia is rich: it can afford to employ several thousand engineers and scientists on the construction of a military aircraft that is specially needed, while Junkers in Germany built the same plane with the aid of fifty skilled engineers.

Russians are perhaps no less skilled than Germans or Japanese, but it really does require a huge "safety margin" to build anything under Soviet conditions, because of all the time that specialists have to spend running from one office to another and obtaining the necessary clearances. Red tape is an inherent feature of the Russian system, and Russians make up for the time it wastes by multiplying the number of those involved. This means that scientists and engineers find themselves doing the work of technicians and craftsmen, but what of that? There are plenty of specialists—too many. And so it is quite natural in Russia to find a setup which would be thought ludicrously inefficient in any Western or civilized country.

At the same time, there is a kind of logic and even political wisdom in the system, with all its red tape and the multiplicity of authorities that have to be consulted. While on the one hand creative thought is strong and valuable by reason of its unexpectedness, the party and state bureaucracy is so organized that it can at any moment put a stop to any initiative it does not care for. This is a normal defensive reaction of the totalitarian state to surprises of any kind. Although much is spoken and written about efficiency, the powers that be do not really mind the fact that Soviet science is inefficient, impoverished, and slow. All that concerns the party apparatus is that science should not escape from the party's guiding hand for a single moment. The hordes of officials who keep a watch on scientists and prevent their working properly know what they are doing; there is nothing unsystematic or thoughtless about the control they exercise.

Thus the enormous number of superfluous scientists is not an accident: they are necessary as a reserve to make up for the "peculiarities" of Russian economy and administration. The authorities know, of course, that mass-produced scientists are no better in quality than other mass-produced articles, but this does not perturb them either. A run-of-the-mill scientific worker, a man of average knowledge and ability, can easily be replaced by another mediocrity; whereas when it comes to talent and personality, there is bound to be trouble. The more easily spare parts can be replaced, the greater the reliability of the machine; and this applies equally to the machinery of Soviet science.

The inflation of numbers has another great advantage from the regime's point of view. Small laboratories and institutes, in which personal originality and initiative came to the fore, are being replaced by huge institutions, as the press and radio propagate the myth that genuine results can nowadays only be obtained by large-scale operations. Research institutes are set up with a staff of one, two, or three thousand. Talking to employees of these "science factories," I have often heard them complain that they lose all sense of personal significance and even cease to believe in their own abilities. They are beset by a feeling of dependence and insecurity; they know themselves to be replaceable and are increasingly intimidated by their superiors. A summons to the director's office fills them with alarm, as does each routine review of their qualifications. The sense of inadequacy, real or fancied, makes them increasingly apathetic and indifferent.

These psychological effects of large-scale research are apparently also felt in Western countries. But in the West a gifted scientist can always transfer to another laboratory. In Russia it is the most gifted and creative researchers who feel the weight of oppression most heavily. A man of more than average ability who wants to try his skill in a different field, or to find something more interesting than his present work, cannot change his job at will. The heads of universities and research institutes are closely linked by ties of common interest, and it is an easy matter for them to bar the door to a researcher who appears too clever or too independent. An instance of how things work was related to me not long ago by Dr. R., a head of section in one of the Moscow academic institutes.

A young researcher on R.'s staff wanted to transfer to a better-paying job in a neighboring institute. But as soon as he applied, Dr. R. received a phone call from his opposite number: "So-and-so wants to leave your staff. Is that okay with you?" In cases like this, people of the same rank, working in the same field and united by party and academic ties, do not need long explanations. The present employer does not have to say anything to the discredit of the man who wants to leave: if he simply replies, "Yes, he may go if he likes," the official at the other end will conclude that the researcher is either incompetent or overly brilliant, and in any case has done something to annoy his chiefs. So the official won't want to take him on or even read his application papers. After a word or two has been spoken at top level, the young scientist will find himself wandering helplessly for months from one institute to another, floundering in the meshes of invisible telephone calls and unable to find work despite his skill and creative zeal.

A young man who has once undergone this kind of trauma will realize the simple truth that he cannot change jobs himself and that any improvement in his position must come from the authorities: it is his business to respect and obey them and to serve without a murmur, wherever they may direct. Anyone who tries to assert his own preferences or defend his rights (heaven forbid!) may be barred from scientific work altogether. This can easily be accomplished, given the enormous reserve and the centralized control of research institutions throughout the country. Among my friends and acquaintances are several who have been permanently excluded from a scientific or academic career in this way, on account of their political or religious views or Jewish origin. Biologists, economists, or specialists in mathematical linguistics or philology, they are now working as carpenters or joiners or are sitting at home in idleness, unable to get back into any kind of creative work.

Needless to say, in such circumstances a researcher learns from his earliest years to conceal his views, feelings, and abilities. Any kind of brilliance is specially dangerous, as it may arouse suspicion or hostility on the part of his superiors. The Nobel prize winner James Dewey Watson, in his book *The Double Helix,* relates that he once arrived in Paris for a scientific congress and, finding he had forgotten to bring any long trousers, delivered his lecture in walking shorts. I am quite certain that none of the many Soviet doctors and candidates of science whom I have met would ever have done such a thing—not because of natural shyness, but because it is out of the question for a Soviet scholar to do anything suggestive of an independent attitude, to be in any way different from the rest. To distinguish oneself is a punishable offense—such is the alpha and omega of good behavior for any of the million rank-and-file Soviet scientists.

Perhaps the most original and independent of my acquaintances in Moscow is the biophysicist Aleksandr P. A man of middle age, he occupies a modest academic position but enjoys sharing ideas with young people. Although attendance at his lectures is optional, the hall is generally full. One day, however, he returned in a gloomy mood: a student had come up to him during the intermission and delivered a monologue along the following lines.

"Everything you tell us is most interesting—so much so that my friends and I cannot wait to explore it for ourselves, to set up our own experiments and observations. We even think we might be able to add something to your ideas on the one hand, and correct them on the other. But where can we find a laboratory to carry on the experiments? It will be at least ten or fifteen years before the luckiest of us is master of his own time and free to conduct experiments of his own. The other day we were doing practical work at an academy institute and they warned us: 'If you want to be taken on here, you must leave your own scientific ideas outside. Only an academician is allowed to have new ideas—or at the very least a doctor of science or a laboratory head.' If that's so, what is the use of all the brilliant ideas our lecturers throw at us? Why call us when there's nowhere to go?"

"To be honest, there was nothing I could reply," said my scientist friend. "The young man was quite right: his career will be like that—unless, of course, he helps himself by methods that have nothing to do with science."

My friend did not specify what these methods might be, but anyone who has been in contact with Soviet students for years would know what he meant. When I repeated the young man's *cri de coeur* to two lecturers at Leningrad University, they nodded understandingly. Yes, they knew the situation well. The days were long past when universities and scientific institutes operated like communicating vessels to ensure the free flow of creative minds and research skills to where they were most needed. Nowadays, when the university, party, and Komsomol organizations decide on a young graduate's future, the last thing they generally look for is creativity. They are not interested in the young man's culture or intellect, his attachment to science or any scientific work he may have done in student groups. What the party committee wants to know is whether he was active as a Komsomol, took part in meetings, helped to run a wall newspaper (a handwritten or typed newspaper pasted on the walls of organizations and institutions), or led political discussions. If he did all these things, he has their permission to be a scientist. I may add that for those who study in Leningrad or Moscow, it is also important to have the right kind of residence permit. A citizen of Tambov or Pskov has less chance of a scientific career than one whose parents live on the Nevsky Prospekt or Arbat Street. This is a police and not a party regulation, but it also goes a long way toward deciding which of Russia's young men and women are to follow in the footsteps of Lomonosov, Lavoisier, or Enrico Fermi.

No secret is made of the fact that political reliability and activity are the main virtues required of future scientists. Old-fashioned professors may still grumble at this fact, but there is no quarreling with the official viewpoint. Some years ago, Yoakhim Romanovich Petrov, an eminent pathophysiologist in Leningrad and a member of the Academy of Medical Sciences, sent me a copy of the *Medical Gazette* with a series of articles under the general title "Who Can Become a Scientist?" Professor Petrov was much interested in this topic, having himself created a flourishing school of thirty-five doctors and fifty-three candidates of science. He was, moreover, a peasant's son, who had begun as a medical orderly and risen to academic rank by sheer ability and industry.[2] In the *Gazette,* he had put a question mark against an article which bore the subheading "Above All, a Strong Sense of Ideology." The article has survived, and it is not hard to see what Petrov found upsetting in it.

> The Soviet scientist . . . must not only defend the principles of Soviet science, but must be able to demonstrate its superiority over bourgeois science. This is specially important at the present time, when international links are developing so rapidly. In choosing staff for research institutes and places of higher learning it is essential to pay closer attention not only to their professional abilities but also to their political maturity.[3]

That was in 1968, but even today any number of equally unregenerate articles could be quoted. The criterion of loyalty as opposed to talent is official doctrine and has been dear to the authorities' hearts for decades. "Ideological sense" is a cornerstone of the system which makes science the handmaiden of politics.

The aspiring scientist is exposed to temptation from an early stage. In the words of G., a teacher of literature at Novosibirsk University:

> In their first and second year, students are fairly unrestrained, expressing independent ideas to their teachers and to one another. On walking tours or at evening parties, you may hear a song or a joke that is suspect from the ideological point of view. The freshmen talk

a lot about injustice and are indignant when their rights are infringed. But in the third year, they seem to undergo a radical change. No more "immature" songs, no arguments with the staff about matters of principle—it's as if they'd been softened up. They hang on their teachers' words, and start taking an active part in Komsomol meetings. That's because in the third year they have to choose a supervisor for their first degree, and to a great extent their future depends on him. He can recommend them for postgraduate work, or block the path of someone he doesn't like. This is the first spiritual test that students go through; not all of them are corrupted, some retain their integrity right through, but in that case they generally don't get to be scientists.

This informant drew my attention to the fact that researchers with higher professional and moral qualities are usually to be found in disciplines that are less pervaded by ideology. It may also be said, as a rule, that the results attained in these branches of science are better than in others. In mathematics, physics, chemistry, and even biology, the party mentors of aspiring scientists are less insistent in their ideological demands. Able mathematicians and physicists recommended by their professors manage to slip through the party net. But a young geographer who wants to study the geography of foreign countries is subjected to the most watchful scrutiny. His ideological "baggage" is turned inside out as though in a customs office, and if anything in the slightest degree untoward is discovered, he can write *finis* to a scientific career. Historians, philosophers, economists, and writers are examined as though under the most powerful microscope.

One advantage, of course, in having a scientific force a million strong is that, however intensively they are screened, some of the most talented and enthusiastic will nevertheless get through the net. And they are needed—after all, someone must do the work. They do not, however, make up a high percentage of the total, as even the official press admits.[4] Soviet sociologists estimate that more than half the country's scientific achievement is the work of 10 percent of its scholars. My own observation would suggest that a considerably smaller number of workers is capable of original scientific synthesis. However great or small the number may be, what is certain is that these original thinkers have a harder time than any of their scientific colleagues.

Some time ago I heard several lectures by V. V. Kovanov, vice-president of the Academy of Medical Sciences of the USSR, and also read his articles in learned journals about an important discovery, that of ischemic toxin. This is a substance which accumulates in the tissues of a live organism which is deprived of oxygen for any length of time; its presence brings about a severe, sometimes fatal shock condition in people who have been crushed in a mine shaft (the "crush syndrome"). The toxin has also had fatal effects on patients who were undergoing surgery for the regrafting of a severed hand or foot. Ischemic toxin represents one of the most crucial problems in modern pathophysiology and surgery, and the medical profession placed great hopes in the fact that it had been isolated and studied.

Unfortunately, while the discovery was undoubtedly a valuable and hopeful one, Professor Kovanov had had nothing personally to do with it: he was only the head of the laboratory where it had taken place. He learned about it only after the work had been done, with much difficulty, by two young scientists, Tatyana Oksman and Mikhail Dalin. This did not prevent him from giving talks in which he claimed the credit, and publishing articles written by the same Oksman and Dalin over his own signature.

For many years I have known all three figures in this story.[5] Years ago Professor Kovanov, who already held a high administrative post in the scientific world, prevented the talented Dalin from studying for a higher degree and refused to let the equally talented Oksman work in his laboratory. The reason? They were Jews. Later, however, feeling the need to add to his scientific laurels, he allowed them both to work for him. Not that Kovanov was himself a villain—if it had not been for his high position, his laboratory, and the luster of his name, the two young scientists would never have had a chance to show that they had discovered something useful. If you asked them what they think of their chief, they would certainly reply that they are grateful to him—he gave them the chance to be scientists, when he could have withheld it.

They showed their gratitude in a practical way. In March 1974 I went to see Tatyana Oksman, who was at home with a broken leg in a cast. I asked if she would be immobilized for long. "I ought to stay like this for two weeks," she replied, "but in three days' time I must go to Leningrad."

"What's the hurry?" I asked.

"Well, there's to be a session of the Academy of Medical Sciences, and Professor Kovanov is reading a paper on ischemic toxin. I wrote the paper myself, so there's no problem there, but there will also be questions." Tatyana knew that Kovanov wouldn't be able to answer the questions, as he had only a vague grasp of the subject, and the danger was that the audience would not take the discovery seriously. Tatyana could not bear to let this happen, so she was going to Leningrad to help out—on one foot, if necessary.

Did she mind that someone else was reading a paper about her own discovery? Not at all—she was used to it. She knew that something of the kind happens every day in the hundreds of Soviet institutes and laboratories. Such is the way of the scientific world, and it is silly to get indignant about it. Sometimes she would joke, with a touch of bitterness, about the "division of labor," but as a rule she tried not to think about it—it was simply a fact of Soviet life.

Tatyana Oksman concealed her humiliation from everyone, even herself. It was the price she paid for being able to do the work she loved, and to advance the cause of science. Her boss was not as bad a fellow as many others. To quote Pushkin's description of a country squire, it might be said that he

> Altered the yoke of old taxation
> From toil exacted, for light rent

Meanwhile Oksman and Dalin are doing work in which they believe, chosen and planned by themselves. Within the laboratory they are their own masters. The boss seldom troubles to visit the place—he has plenty of other duties as a high official of the Medical Academy and a professor at the Medical Institute. But when it produces results, he demands his rightful share—or rather he does not have to: it is offered spontaneously and gratefully. After all, he does not take everything, as some chiefs would do, but only just as much as he needs to keep up his academic reputation—and is it such a hardship to write a paper for him to deliver at a session of the Academy, or an article for a learned journal?

"Rent" or "tax" of this sort is the most usual basis of the relationship between senior and junior scientists in the Soviet Union. Without such tributes the administrators of the "million," who are consumers and not producers of science, simply could not exist:

they could not perform their functions, get promotions, be awarded state prizes, deliver papers to the Academy, or attend international congresses.

Within this system, which prevails throughout the country, each individual boss operates after his own fashion: the "tax" exacted by Kovanov is mild indeed compared with the forced-labor regime of, for example, Yuri Anatolyevich Ovchinnikov—a vice-president of the Academy of Sciences who is also director of the Academy's Institute of the Chemistry of Natural Compounds, chairman of two academic committees on important biochemical problems, professor of biochemistry at Moscow University, head of a large biochemical laboratory at the Science City of Pushchino near Moscow, and chief editor of the journal *Bioorganic Chemistry*. Ovchinnikov, who is in his forties, enjoys a complete monopoly of biochemical studies, and his scientific domains are subjected to a quasi-military regime.[6]

The main purpose of Ovchinnikov's research is to impress his superiors and foreign colleagues. He chooses a subject which is specially promising from the propaganda point of view, such as the chemical structure of one of the proteins (five hundred of these having already been analyzed, he will tackle the five hundred and first). Using his high administrative rank to commandeer equipment and chemicals which are in short supply, he instructs his staff to lay off all other investigations and concentrate on the selected "product." The laboratory turns into a kind of anthill: each individual works at his own narrow task, ignorant of the common aim, with no idea what his neighbors are doing or what purpose their efforts are meant to serve. All the threads lead to Ovchinnikov himself: he alone knows the object in view, he can tell when successes are achieved or mistakes made, and he alone takes the credit for the final outcome. Thanks to these shock tactics he had, at the age of thirty-six, become an academician.

No one would envy the anonymous, serflike mass of assistants working for Ovchinnikov: they are harried and bullied by their boss and, far worse, the mass-production system deprives them of all creative pleasure in their work. But this is the last thing Ovchinnikov cares about. He says openly that he doesn't need anyone's brains—his own is good enough; all he needs to achieve his purpose is a sufficient number of hands.

Ovchinnikov's method is typical of Soviet science as a whole, and the party's leaders point to him as a paragon of energy and organization. He is generally expected to be named president of the Academy in 1980; for no one has understood better than he what splendid prizes are in store—no, not for science, but for anyone who contrives to organize scientific slave-labor on a large enough scale.

The independent researcher, freely following his own inspiration, is practically extinct in the Soviet Union.[7] A scientist can be only a master or a servant. Young people with ideas are depressed and discouraged, not so much by the fact that others take credit for their work, as by the sharp and increasingly rigid division between scientists and administrators, between the serfs and those who enjoy the fruits of their labors. Anyone who throws in his lot with the "million" soon discovers that it is no use being a scientist unless you have authority on your side. Before you can put forward ideas on linguistics or pediatrics, crystallography or astronomy, you must first show that you are capable of running a department or an institute, acting as secretary of a party organization, or, at the very least, commanding a laboratory. Otherwise no one will bother to talk to you. But in this struggle for administrative positions, the genuine scientist is at a disadvantage: it is much easier for a bureaucrat, appointed from above, to gain control of the apparatus than for a scientist, however talented, working his way up from below. Moreover, the bureaucrat has a choice of occupations: instead of heading a research institute, he can always be given a theater to run, or a laundry system—or even the Writers' Union. A physicist or botanist has no such options; but to be accepted as a scientist, he must have an official position as well. That position acts as a "hot line" enabling him to talk to his superiors; it provides him with money to buy equipment, opportunities to travel at public expense (including official trips abroad), a larger establishment of his own, and a chance to publish his work in scientific journals. In short, an administrative job is a passport without which he can get nowhere in his chosen field. Those are the rules of the game. Thus, if you are one of the 10 percent who are capable of original work, and if you do not want to be a serf all your life, you must join the rat race and compete for jobs with bureaucrats who represent the opposite of all you stand for.

My notebooks are full of stories illustrating how the paradox works out, with comic or tragic results according to the character and principles of the individual scientist.

Professor G. B. is a surgeon, aged forty, a dynamic character who loves his profession and is keen to advance in it. He works in the small town of Yoshkar-Ola, which many Moscow intellectuals might find hard to place but which is the capital of the Mari Autonomous Republic on the middle Volga. A university has recently been opened, and Professor B. lectures at the medical faculty. However, he wishes to pursue scientific research, and, in particular, to develop and test drugs which promote the regeneration of tissues after surgery. (This was not originally his own idea, but it is an important and serious problem.) He has published several articles on the subject, and now needs a laboratory to supplement the work done in his clinic. He is prepared to run such a laboratory without any additional salary.

First Professor B. applied to the party district committee, where he gave a carefully simplified account of his idea to the Mari official in charge of cultural and scientific matters. The official agreed that it would certainly be better if wounds healed immediately instead of taking three months, as had been his own experience when head of the local collective farm. He gave the professor a letter of approval, without which the application would have been useless. But this was only a beginning: the laboratory could not be set up without permission from Moscow. Professor B. went to Moscow—he went, in fact, as many as eight times, by the end of which period he had a whole file of letters endorsing his plan, including one from the president of the Academy of Medical Sciences and one from the chairman of the Academic Council of the Ministry of Health of the USSR. He needed only two signatures more; those of the Minister of Public Health of the Russian SFSR and the All-Union Minister of Higher Education. However, one cannot go straight to the top: the proposal had first to be reported on at each level in the two ministries. After a day spent running from one office to another, Professor B. told me his story:

I managed to get an interview with the head of the scientific department. I gave him a Mari souvenir—a wooden ladle—and a bottle of good brandy. He listened politely, called in an inspector—a youngish woman—and told her to "do a brief" for the collegium of the Ministry. I gave the inspector a box of chocolates, and she said she would have the papers ready for the next meeting of the collegium. But after that it will have to go to the collegium of the Ministry of Higher Education. What gifts can I take them? I might present the deputy minister with some good books out of my library —but I don't know, I've already spent quite a lot as it is. . . .

That was in the fall of 1973. By the spring of 1976, Professor B. still had not collected all the authorizations he needed to open a laboratory for the study of regenerative processes at the Mari university. Perhaps in a year or two he may succeed.

Another friend had an even more difficult problem. Stanislav G., a candidate in biological studies, is one of the "10 percent", a man with ideas of his own. He is about forty. He has evolved new methods of isolating genes and he is employed at the Institute of Genetics of the Academy of Sciences, directed by Academician N. P. Dubinin. Foreseeing that Stanislav (Slava for short) would produce interesting results and that the credit would redound to himself, Dubinin allowed his assistant somewhat greater than average freedom. Slava was a hardworking chap, prepared to spend day and night in his laboratory.

But hard work and talent are not enough in modern genetics— you need a staff and equipment. Slava was not even a laboratory head. He had only a handful of assistants and the work was taking a long time. He went to the director, who listened sympathetically and offered him a piece of good advice. Why did Slava not get the present head of his laboratory thrown out, so that he could take his place? That would solve all his problems; then he could set the whole laboratory to work on his project.

This advice was thoroughly typical of Dubinin, but it did not appeal to his young assistant, who replied that it would be against his moral principles to supplant a colleague. "I understand you're a believer—a Christian?" Dubinin asked. Yes, this was so: Slava had not concealed the fact, although it was not without danger to admit it in a scientific institution. Dubinin looked at him with a trace of contempt. "Well, it's your affair," he muttered.

Slava's attitude toward religion had been duly reported to the director, and one of these days it might be used against him. It was not yet time for that, but meanwhile the academician had received further proof of the harm that religion can do a man's prospects.

Slava went back to his laboratory and worked even harder. He had been a laboratory head in the past and was not anxious to be one again: it meant spending most of his time scribbling reports and sitting at meetings, pencil-pushing instead of getting on with experiments. Of course others did the work meanwhile, but Slava had no wish to be a parasite. But what was he to do? he asked himself again and again. Years went by, the work was no further advanced, and the director's words rang in his ears: "Why don't you get rid of so-and-so? You'd be doing yourself a good turn, and science would benefit as well."

The advancement of science is one of the favorite slogans of our age. A man who succeeds in ousting his superior and taking charge of the laboratory, or one who exploits his staff so that he can be the first to proclaim a new discovery—all who practice these and other tricks justify them by talking about the benefit to science. As for party committees within academic institutions, the phrase is never off their lips.

In the Leningrad Physicotechnical Institute named for Feydor Yoffe, there is a candidate in physics, a quiet, industrious character whom we shall call X. He is not a man of outstanding gifts, but he is also without pretense. If he is called on to attend a demonstration or to pick potatoes, he obeys without question. He does not seem to play dirty tricks on anyone—he's a bit of a coward, certainly, but who isn't nowadays? One fine day he is called in by the party committee, and a friendly chat ensues.

"We've been keeping an eye on you for some time. You're a promising young scientist, you show self-discipline, and your colleagues respect you. Why don't you put in for membership in the party? We'll support you."

"Well, you know, I—I never thought of such a thing," stammers X. What does he want with politics? Physics is all he cares about. He doesn't say that out loud, of course, but memory supplies him with a saving formula: "I don't feel mature enough," he blurts out, looking hopefully at the party secretary. Will it work? After all, ideological maturity is so important. . . .

But the secretary brushes the excuse aside. "Never mind that, you'll get mature as you go along. What matters is to do something for science. You're going to get a doctor's degree soon, aren't you? Well, after that you'll need assistants, equipment, grants. . . . How will you get them if you aren't a party member? Everything in Soviet science goes through the party organization: degrees, funds, jobs, equipment—everything. If you join the party, we'll recommend that you get a lab of your own, and then you'll have all the staff and money you want—you can organize everything for the greater good of science. So don't be difficult—there's a good fellow—but just read the rules and the party program and come along to us with your application."

X returns to the lab and tells his friends what has happened. He doesn't want to join the party, he wants to be a physicist—listening to speeches and taking the floor at meetings doesn't attract him at all. It's all right when there is nothing special going on, but supposing it's like 1968, when party members had to say how splendid it was that Russian tanks went into Czechoslovakia. Or 1973, when they had to sign indignation letters against Sakharov? As a non-party-member you can sit tight, but party discipline's another matter. . . .

But X's friends take a different view. "Go on," they cry with one voice. "Go on, old fellow, and join the party—it'll be better for you and for us. You'll be head of the lab. We know you; we can work together all right. But if you won't do it, they'll bring in some jerk from outside who'll make our lives miserable. Go on, don't give it a second thought! It's for the good of science!"

X looks at them dolefully. To tell the truth, they are not very happy about urging him on in this way. It's like persuading a man to have an amputation—just a slight moral amputation, after which everyone will feel better and more comfortable.

Then X tries his last line of escape: "All right. I don't mind as far as I'm concerned. But what about you, why don't you all join the party?"

His friends burst out laughing. "That's a good one—why should we? They won't put us in charge of labs if we do—and we're still a long way from getting our doctors' degrees."[8]

In the world of the scientific million, power is measured first and foremost in terms of laboratory equipment. The modern researcher needs apparatus almost as much as he does ideas. Indeed, for immediate purposes a good new piece of equipment is even more important than a new idea, since with the equipment you can do some kind of work in any case, but ideas are little or no use on their own. However, most Soviet laboratories are very badly equipped, except those working for the Army or run by academicians who are also high-powered administrators. New apparatus is allocated strictly according to rule: all the best items go to military suppliers, and after them to the big bosses, while others are given rubbish or nothing at all. Every now and then, as a result of the chronic deficit, squabbles break out within ministries or academies over this or that piece of equipment. The Soviet Union produces only a small quantity of modern electronic and optical instruments, of greatly inferior quality to those made abroad. The latter have to be bought with foreign currency and are assigned only to the top echelons of the scientific administration.

The fight for equipment is the most constant feature of Soviet scientific life. Back in the 1930s, there was a joke to the effect that all the institutes in the country were terrified of losing their equipment to Bakh. A. N. Bakh (1854–1946), a biochemist and director of the Institute of Applied Chemistry, enjoyed the Kremlin's favor because of his devotion to the party line and his fantastic plans for "chemicalizing the whole country,"[9] and he did in fact commandeer any piece of special apparatus that took his fancy.

By the mid-1950s, when more equipment was being purchased abroad, the fight grew still more intense. To be on the safe side, high officials of the ministries and academies got items that they did not need and did not know how to use. Apparatus purchased by the state from the United States, Sweden, Britain, or Japan was treated as the property of individuals, and very valuable property too: the staff of the scientist to whom it "belonged" could carry out important and highly competitive work, and of course he himself shared in the credit. Moreover, he could lend "his" equipment to others—not in return for money, but to share the credit for their discoveries. In short, the scientific bosses discerned the merits of the new equipment even before it was tested by their experts.

The importance of this matter was made clear to me, when I was a young journalist in the mid-fifties, by Boris Petrovsky, a surgeon who is now an academician and Minister of Health of the USSR. At that time he was a professor and head of a clinic at the Second Medical Institute in Moscow. I visited the clinic several times, as I was writing my first book about medical specialists. I did not get along very well with Petrovsky, who made it clear from the outset that he expected to be the sole hero of my book, and even sent his assistant to explain to me what an important figure he was. I made friends with the other surgeons, however, and was allowed to witness operations and even sit through the night watches to get a better idea of the problems of tending patients in critical condition. One day, however, when I turned up at the clinic to sit in on the usual morning conference, I could hardly recognize my friends: the men looked gloomy, and the women had evidently been crying. The atmosphere was funereal, and nobody would tell me what had happened.

I discovered, however, that Petrovsky had been offered—and accepted—a more distinguished post as head of the First Medical Institute. Any professional man has the right to choose his place of work, but Petrovsky had done more: he had given orders that all the diagnostic equipment that his pupils and assistants had been using was to be unbolted from its concrete foundations and removed to his new clinic. The operation was carried out in such a hurry and in such secrecy that it was not until the surgeons saw the empty laboratories that they realized Petrovsky had decamped and taken the equipment with him. In many cases this meant that they were deprived of their doctoral theses as well, since their research into the diagnosis of heart ailments was based on data obtained with the unique apparatus in question.

Many a prosperous academic or professorial career has been based on the ability to get hold of valuable apparatus and use it to one's best advantage. The staff of a research institute at Chernogolovka near Moscow told me admiring tales of their former director, not so much for his scientific ability as for his extraordinary skill at acquiring equipment. He had an unrivaled knack of getting the procurement agency on his side, intercepting supplies earmarked for other laboratories, ingratiating himself with the grants committee, and similar ploys, to the benefit of his staff as well as himself. He was transferred long ago to another institute, but his former assistants still remember him with respect and gratitude. Without such skills as his, research in many a Soviet institution would simply come to a stop.

Yet Chernogolovka is by no means a backwater. A small town in the midst of pinewoods thirty miles from the capital, it contains several academy institutions with a high reputation, including a branch of the Institute of Chemical Physics headed by the Nobel prize winner N. N. Semenov. In December 1975 I spent some days there as the guest of Gerz Ilyich Likhtenshtein, a gifted researcher about whom I was then writing. Although he was a Jew, he had managed to become head of a laboratory some ten years earlier and had been able to do some very interesting work in chemistry, physics, and biology. Some of his discoveries had won international recognition: in particular a method of using radicals as markers in biological systems, and the study of a process of considerable economic importance, the "soft" fixation of atmospheric nitrogen by the fermenting of soil bacteria. He had published a book in the United States in 1975, and the thoroughness and originality of his work were admired by scientists in many countries. He showed me appreciative letters from eminent researchers throughout the world.

I was naturally interested to learn how Professor Likhtenshtein had come upon his discoveries, but he spent a great deal of time telling me how he and his staff had gotten over the lack of equipment and chemical reagents, which had held up his work for several years. The shortage of equipment and consequent failure of experiments had at times threatened not only the realization of his ideas but his actual livelihood as well. However, being an optimist by nature he maintained that this state of affairs, which had lasted as long as his own career, had done much to stimulate his inventive faculty. As he said with wry humor, "When you have no apparatus you really become an original scientist."

Likhtenshtein showed me a radiospectrometer of electroparamagnetic resonance, an instrument of great importance in the study of the kinetics of chemical reactions. This device had been acquired for the Soviet Union about twenty years earlier by Dr. Zavoysky of Kazan, and had been produced in Russian factories for the past fifteen years, but during that time no modernizations or improvements had been introduced. Western firms were producing a much perfected and more convenient version. "Unfortunately," said Professor Likhtenshtein, "I have no gift at all for getting hold of equipment. It is easier for me to think up a new method or a new contraption than to wheedle the authorities for a new piece of apparatus—even some old stuff that nobody wants." He confessed that he felt bitter and depressed when he read of the experiments conducted by his colleagues in the United States and elsewhere, at a level of technical perfection that he could not dream of equaling. Although his ideas are no less original and important than those of foreign scientists, and perhaps are more so, he is beset by a feeling of inferiority due to the miserable conditions in which he operates. Young candidates of science in his laboratory do not feel the situation so keenly, but they too realize that the shortage of "hardware," as they call it, cramps their style and originality: they have even composed little satirical songs on the subject.

What with bureaucratization, serf-labor methods, and the general dearth of equipment, the five thousand Soviet research institutions are, on the average, about as efficient as Denis Papin's seventeenth-century pressure cooker. Articles in learned journals take months to appear, experiments go on for years, and the practical application of even a major discovery may take decades. A member of a research institute supplying army equipment admitted to me that for years he and his colleagues had been used to seeing their latest model in a finished state just as its design was becoming obsolete. And many stories could be told of the failure of Soviet science to affect agriculture, the very small range of new drugs produced in the Soviet Union, and the minimal effect of scientific discovery on such products as light automobiles, clothing, housing, and foodstuffs.

What do Russian scientists themselves think about it?

It is easiest for cynics. Abel Gezevich Aganbegyan (born 1932), an academician and director of the Institute of Economics and Industrial Organization at Novosibirsk, was once asked whether the USSR could overtake the United States in science and economic development. He replied that if that should ever happen, the Soviet Union would have to stop and let the United States get ahead again, since if we did not have the Americans in front of us we would not know which way to go. In Aganbegyan's opinion, the achievements of Soviet science are usually a function of those of the United States: as presently constituted, Russia's scientific community does not and cannot produce any fundamentally new and original results.

Dr. R., who also occupies a high position in an Academy institution, is not a cynic like Aganbegyan, and is grieved by the very low productivity of Soviet science. But he, too, is convinced that Russia has lost the race with the West in this field and in humanitarian studies as well, since scholars and scientists cannot work properly in an atmosphere of forced conformity and bureaucratic supervision. In a confidential conversation, Dr. R. described his sad conclusions:

It was painful to me to feel at odds with my own country. But that is in the past. I don't wish to emigrate; on the contrary, I wish to study this country as a virologist studies a virus, or an oncologist a tumorous cell. The trouble is not so much that the government and institutions are at fault, as that we scholars and intellectuals have nothing concrete to offer. This applies not only to chemistry or engineering, but even more to history, philosophy, and ethics.

Views like those of Aganbegyan or Dr. R. are only a ripple on the surface, but the million rank-and-file scientists have their own grounds for misgivings and discontent. The average senior or junior research officer is quite intelligent enough to see how desperately slowly the work is progressing, and how many valuable ideas are going to waste. He knows that the Japanese protein analyzer is far better than the Soviet one, and that if you value your patient's health you will use a British and not a Soviet contrast preparation for X-raying blood vessels. But these things worry him much less than whether his appointment will be confirmed for another term, whether his doctoral thesis will be approved, and whether he can get the director to obtain a special kind of photographic film. "Social problems, philosophy, morality—I pushed all that to one side years ago," said Dr. Oleg Ivanovich Kirillov of Vladivostok. "I haven't time for it, and anyhow it is bad for one's work." Kirillov, a pathophysiologist and pharmacologist, has pursued original research on the basis of Hans Selye's theory of stress, but, unlike his Canadian colleague, he shrinks from discussing or even thinking about its social aspects: he has known from his earliest years that doing that would get him into trouble.[10] The backwardness of Soviet science is a social problem too, and in Dr. Kirillov's opinion that is sufficient reason for ignoring it; better concentrate on your work and avoid taking risks. I believe this view is shared by hundreds of workers in all branches of science.

The rank-and-file scientist cannot give up his subject entirely without losing his livelihood, but neither can he work under existing conditions to advance the cause of science. The most he can do is to go through the motions of performing research that will look good from a bookkeeping point of view. Of course there are still quite a few who are determined to pursue genuine research, to invent and discover at all costs; but the system is heavily weighted against them.

The poor performance of mass-produced scientists in recent years has even been mentioned, though very cautiously, in the official Soviet press. B. N. Volgin, a candidate of technical sciences, wrote in 1971: "All is not well with the efficiency of our scientific work. As scientists get more numerous, the level of originality declines; this is a phenomenon that can be observed everywhere."[11] Volgin even goes so far as to speak of a "provincial and lackadaisical atmosphere" pervading Soviet research institutes. Soviet scientific journals traditionally avoid social problems, but the matter has been ventilated in the pages of the *Literary Gazette,* where shortcomings have been ascribed to insufficient pay and defects of internal organization. The psychology of the scientist and the ethical attitudes of the "million" continue to be forbidden subjects.

On the other hand, as the "lackadaisical" spirit spreads to one laboratory after another, as research targets fail to be met and government projects come to nothing, the bureaucracy is at hand with remedies. An example of these is a form, printed in a fairly large number of copies and reading as follows:

Department of Automatic Control System, Sklifasovsky
First Aid Research Institute

Technical assignment: _____
Officer responsible for execution: _____

The task of _____, assigned to you with a completion date of _____, has not been finished by the due date. I draw your attention to the inadmissibility of this attitude to the tasks laid upon you, and I expect the assignment to be fully completed by _____. No excuses for further delay will be accepted.

Head constructor of automatic control
system, central Public Health Administration:
I. Beskrovny

Dr. Beskrovny probably does not himself believe in the magic efficacy of this multigraphed threat, but he is himself one of the million and is playing the game by the rules. He fills in the name, dates, and other details, sends the form to the offender, and ticks off an item in his diary: "X.Y. warned." That is what he is there for.

CHAPTER 4

A SECRET INSIDE A MYSTERY

"'Fellow citizens!' he began in passionate tones, but, as his speech was secret, of course no one could hear it."*

"As is well known, secret information is more reliable than non-secret."†

On a December morning in 1975 the Kutuzov Prospekt in Moscow was suddenly filled with the wail of sirens, as fire engines converged from all directions on a huge administrative building near the Kutuzov metro station. The fire was on the sixth floor, and smoke was pouring out of the windows. Passersby looked on with interest as no fewer than twelve engines appeared. They would have been still more intrigued had they been in the lobby of the building when a team of firemen, wearing masks, rushed in and were confronted by a group of security guards with green cap bands, who demanded that they show their passes.

"What do you mean, passes?" retorted the fire chief. "There's a fire in the building. Get out of the way!" He tried to push the nearest guard aside so that he and his team could mount the stairway, but the guard stood fast.

"Stay where you are!" cried the chief guard, as he and his men unbuttoned their holsters. "Fire or no fire, no one gets in without a pass. This is a high-security area."

"Let us through, damn you!" was the frenzied answer. "Can't you understand, there are people burning to death up there!"

"Let them burn," said the other, unmoved. "You can't go up without a pass, and that's all there is to it."

* M. E. Saltykov-Shchedrin, *Complete Works,* Vol. 8, p. 302.
† Ibid., Vol. 3, p. 267.

So the firemen turned around and retreated, with the security men's revolvers still leveled at them.

While this altercation went on in the lobby, the fire on the sixth floor grew more and more intense. Valuable equipment and papers went up in flames, as well as some eighty gallons of kerosene. The only way the firemen could get in was to put up ladders and climb in through the upper windows. The extent of the damage and loss of life was never published, but the security guards were declared to have behaved properly in refusing the firemen access.

This was fully to be expected. Secrecy is the main product of hundreds of Soviet research and development institutes, and may be called the lifeblood of Soviet science. The staff of an institute may turn out inferior products at huge expense, so slowly that the equipment is obsolete before it comes into use, and nobody will turn a hair. Even if the building on Kutuzov Prospekt had burned to the ground, the director and his senior assistants would not have been too severely punished—but God help them if there had been any security breach or "leakage of information."

In order to understand why the Russians guard scientific achievements so closely, one must recall certain basic myths of the closed society. The first is that the citizens of the socialist state are uniquely blessed in that they live in a progressive society, the land of universal prosperity and contentment, while the reactionary world outside looks on it with hatred and envy. The Soviet citizen must therefore be constantly on the watch, keeping his powder dry, his arms at the ready and his safe triple-locked. This myth is associated with such slogans as "mass espionage by foreign intelligence," "our renowned security and counterespionage officers," "the frontier locked and barred," and so forth.

Another myth is that Soviet science is the most advanced in the world, and that Soviet scholars are constantly making wonderful discoveries in all branches of learning—in short, there is plenty that is worth stealing or spying on. Every institute and laboratory must therefore be turned into an impregnable bastion of secrecy, and every scientific unit must be guarded from foreign incursion. In a world divided into two camps, progressive science needs secrecy to defend its conquests—such is the official version.

In itself this theory is barely half a century old, but the roots of

Russian secretiveness lie much deeper, in a tradition that goes back for centuries. As the Marquis de Custine wrote 140 years ago, "in Russia they make a secret of everything." Denis Diderot, who spent some months at the court of Catherine the Great, noticed the same thing 60 years earlier, and in the sixteenth and seventeenth centuries the suspiciousness of the Russian character was noted with amazement by European travelers, such as Schlichting and Olearius. The view that every foreigner is a dangerous spy pervades the ideology of the Russian state and the Russian people. An iron curtain of fear and mistrust divided Russia from the outside world long before "proletarian internationalism" was ever heard of.

When a tradition of this sort is fortified by material interest, the combination is an extremely strong one. The anonymous guard who threatened to shoot down the invading firemen is a symbolic figure. As to his personal interest, it should be borne in mind that a security officer guarding a research institute is much better paid than, for instance, a laboratory assistant without a higher degree, while generals in the security service are paid about as much as professors. It is only natural, therefore, that the guardians of state secrets perform their work zealously.

I once attempted to calculate the total number of such security officers—not rank-and-file guards, but members of what are called First Security Departments. Such departments exist in every university, research institute, and independent laboratory. Most of them are amply staffed, and the three hundred research institutes in Moscow account for a whole division of warriors protecting the nation's secrets. In the whole of Russia they cannot be much fewer in number than those performing the opposite function, that is, members of the scientific information service, of whom there are 100,000.[1]

According to the older generation of scientists, this blanket of security did not descend all of a sudden. Up until World War II secret projects were extremely rare, even in research institutes concerned with engineering, physics, and chemistry. In the State Optical Institute, for instance, there were only two or three such projects per year. Yet today it is quite difficult for an optical specialist, armed with an official letter stating his business, to gain admission to the Institute building, and for anyone else it is impossible. Even after obtaining a pass, which may take hours or days, the visitor is not allowed access to the whole Institute, as special passes are required for admission to floors where top secret work is being carried out.

The advance of secrecy went hand in hand with militarization. After the war, scientists were enrolled throughout the country to work on military projects. Secret laboratories sprang up in almost every university and technical institute. Any new scientific idea of any interest was commandeered for military purposes, with no expense spared. Gradually researchers whose work had no connection with defense problems were drawn into the militarist network: high salaries were offered them as an inducement to work on some narrow aspect of their subject, and their laboratories were subjected "temporarily" to security restrictions. But while it is easy to impose such restrictions, it is next to impossible to get them lifted. Consequently, one scientific unit after another was caught up in a huge spider web of secrecy, and this process is still continuing. Workers in secret laboratories cannot publish articles or take part in official symposia, and they are cut off from scientific contacts of every kind.

On a few occasions in recent years there have been appeals for the relaxation of official secrecy, for instance by the academician and radio engineering specialist A. I. Berg, who exclaimed: "We are stuck fast in secrecy like a fly in treacle. It's impossible to work like this!" But his sensible words were not heeded. The two or three token measures of derestriction since the war amounted to no more than lifting the veil from processes and equipment that had been in existence for thirty years.

At the present time there is scarcely a research institute in the country whose work is entirely in the open. Institutes of apiculture or wild-life husbandry may be an exception, but even their members are forbidden to publish articles or books mentioning a falloff in the honey harvest or the production of animal pelts. Even in Moscow institutes to which foreigners are invited, where academicians grant interviews and do everything to show how free they are, many doors remain firmly closed to the outsider. You may freely visit the second floor of the Institute of Physical Problems of the Academy of Sciences of the USSR, directed by P. L. Kapitsa, but the first floor consists of secret laboratories. A good half of the

Levedev Physical Institute is closed to unauthorized persons, and the same is true of the N. N. Semenov, A. N. Frumkin, A. N. Nesmeyanov Institutes, and many others. But the security men's paradise is, of course, that part of the scientific complex which is concerned with military problems.

The salary in a "classified" institute is a good deal higher than in an "open" one. A junior assistant who has only a candidate's degree, or perhaps not even that, gets tired of living on his meager pay; one day a friend or a former fellow student tells him the address of a secret institute related to his specialty, and he hastens along, documents in hand, to a building with no address plate on its front door. He need not in fact be in such a hurry—his application is bound to be scrutinized for at least three months, and quite possibly for a year and a half. If it is rejected, he will never be told the reason: it may be because he is a Jew or has Jews in his family, which is an absolute bar nowadays, or possibly a neighbor has informed against him.

However, let us suppose his application is accepted. He is next made to sign a paper promising not to breathe a word about his job to anyone at any time—not to his wife, his son, or his dearest friend. He also solemnly undertakes to have no dealings with foreigners—not to make the acquaintance of any, not to invite them to his home, never to travel abroad or write letters to foreign countries, on threat of prosecution under the Penal Code. There is also a penalty—three years in a labor camp—for losing his Institute pass. Having confirmed by his signature that he understands all this, the lucky man obtains the coveted pass and is admitted to the sacred portals.

When he reaches the laboratory he soon finds out that there was not much point in swearing him to secrecy about his job, because (1) a junior research officer works on the construction of a single unit of equipment, in ignorance of the other units with which it is to be combined and the nature of the complete product, and (2) the main task of secret laboratories is to copy models manufactured in the United States.

Nonetheless, the rule of secrecy is rigorously observed. Five men, sitting in the same room and possessing different passes according to the degree of their initiation, are forbidden to discuss technical difficulties and successes among themselves, or even to cast a glance at one another's drawings. In one research institute the security department had forgotten or had not gotten around to issuing a pass to the inventor of the apparatus on which the whole laboratory was working, and for a long time the unfortunate man was not allowed to touch the drawings relating to his own brainchild. Nor was he allowed, on threat of imprisonment, to reproduce them himself, above all not in his own home or anywhere outside the institute.

The fortunate recruit very soon begins to notice that his laboratory colleagues are afraid to say an unnecessary word to one another, and he himself learns to watch his tongue and confine himself to trivialities. Having made some more or less commonplace remark, he will find himself looking intently at the face of the man he is talking to, for fear that he may be an informer. However, as man is a social being and it is natural to want to exchange ideas, our hero tries to talk to his colleagues about art or literature; but he finds to his sorrow that they are not interested, and only want to discuss football and ice hockey. Monotonous work, without a clear purpose and without any creative ardor, has a deadening effect on the mind. The more gifted and thoughtful types seek oblivion in drink or debauchery; but the majority are not distressed and find nothing strange or unpleasant about their position. It is all perfectly normal: they do their work and get paid for it, and even have the opportunity to present a dissertation for a doctor's degree, which entitles them to a double salary. What is there to worry about? The work may be drab and dispiriting, but the lab isn't a place for having fun anyway. . . .

Our hero continues to live in this way for one, five, or ten years. If he is not a party member, he will soon reach his ceiling as far as work is concerned, and any intellectual or material stimulus will have completely gone out of it. Unexpected events may occur, but not of a creative kind. For instance, a research officer was asked by his wife, a doctor, to take the annual report of the maternity home at which she worked to his institute to have it bound. The work was done, but when he tried to leave the premises with the volume in his possession he was stopped by the security guard. It was no use showing the text of the report and pointing out that it was about primiparas and multiparas, abortions, and ruptures of the perineum. The guard merely repeated "Secrets are secrets," and insisted that he get permission from the First Security Department.

Let us say, however, that a ray of light dawns in our hero's drab existence: he has an ingenious technical idea and would like to discuss it with his colleagues. The laboratory head, however, advises him not to make too much of it: there have been no instructions about such ideas, the Americans haven't thought up anything of the kind, and it would be just as well not to get too excited. There being nothing secret about the idea, the researcher asks permission to publish it in an unclassified journal, but after some months' delay he gets the answer from Security that it can only be published in a classified one. By that time he has lost the urge to publish it at all. He surrenders the draft of his article to the First Department and tries to forget his lapse into originality as quickly as possible; both his immediate boss and his wife breathe a sigh of relief at his decision.

Inured to this sort of life after some years, the researcher begins to think of his job merely as a way of receiving, twice a month, a higher salary than he would get elsewhere. Sometimes, after a glass or two of vodka, he will start thinking that his work is a preparation for war and slaughter, and this will make him feel even more disgusted. But such feelings occur less often year by year. In the first place there is no one to share them with—his colleagues prefer not to talk about such things, and if they did they would simply remind him that in an unclassified institute he would only make half his present salary. Thus he ends up resolving the eternal problems of war and peace in routine fashion, with the help of the vodka bottle.

I have not invented this sketch of the researcher's career—everything I have described, and much else besides, has been told me by friends and relatives who are or were employed in classified institutions. Many of those who related such things found nothing repulsive in them, and regarded the excesses of secrecy as perfectly natural. One, a doctor of science in his fifties (who, I may add, once had to hide his candidate's thesis under his belt in order to get it out of the building, even though there was nothing secret about it) said in a melancholy tone: "Of course secrecy is humiliating and stifles the creative impulse, but most researchers in classified institutes are thick-skinned and not very bright. They live in a world of secrecy without suffering or even feeling irked by it, so there's really no need to pity them."

I find it hard to agree completely with this view, since there is at least one point on which members of classified research institutes have strong and deep feelings. However "thick-skinned," however depressed and corrupted they may be, every one of them would like to believe that there is some purpose in his dreary life, that his work is not wasted but is in some way useful to his country. But even this hope is constantly being thwarted. An engineer with a candidate's degree in technology, who had worked for many years in a classified research institute, told me:

"For more than a year we—that's to say a big laboratory, several dozen of us—were working out a complicated system, based on an American design as usual (what the juniors jokingly call a 'translation from the American'). All went well, we were expecting to see the end of our work and get a prize for it any day, when I happened to come across our top secret drawings in a U.S. journal in the Institute library. While we had been taxing our brains, the Americans had taken the system out of production and declassified it. Full of disappointment, I set off with the journal in my hand to find the laboratory chief. On the way I stopped in the corridor to have another look at the published diagram. A security officer who was hanging around pounced on me and asked what I was looking at. 'Diagram number so-and-so.' The man's eyes gleamed as he scented a security breach. 'Who gave you permission to take it out of the lab?' I showed him the cover of the American journal. He gave a horrified start and exclaimed: 'So they've got wind of it!' The poor fellow obviously thought his career was ruined. I had to reassure him and explain that it was we who had gotten wind of it, not the Americans, but unfortunately we were years too late."

Contretemps of this sort are not unusual, and, although my friend described the incident with humor, such occurrences caused him and his colleagues no little distress. A professional fiasco is much more painful to the mass-produced scientist than any moral dilemma. When he discovers that the work of a whole team has been wasted and that millions of rubles have been spent in vain, he feels that the last justification of his existence has been removed, and that he is a wage earner and nothing more.

There is, however, another type of researcher in a classified institute for whom secrecy is a happy invention, opening up the prospect of an endlessly successful career. Such men as these are really at home in a top-secret atmosphere. They seek out suitable institutions to work in, and even sometimes create them for themselves. One such fabulous estabishment came into existence some years ago near Moscow, bearing originally the modest title "Post Office Box No. ——." Its history deserves to be described in detail.

Soon after N. G. Basov and A. M. Prokhorov received the Nobel prize for the construction of quantum generators, the military inquired of the physicists about the possibility of developing a long-range laser weapon. The heads of the military-industrial complex, with billions of rubles at their disposal, were carried away by an idea which had originated in the mind of a Russian science fiction writer in the 1920s: a "superbolide" which would be stationed at a great height thousands of miles from Russia's borders and would intercept foreign missiles, slicing them in two as a butcher slices a sausage. "Very well," replied the scientists, as they always do where the military are concerned, "we will make a laser ray and place it at our country's disposal."

Thereupon the order was given to set up a secret institute near Moscow to produce the superweapon, at a cost of no matter how many millions. Its director, to make things as reliable as possible, was a member of the Science Department of the Central Committee of the Communist Party of the Soviet Union. However, this eminent physicist (he had once taken a candidate's degree) very soon realized that the project was unworkable: superbolides were as much of a fantasy as they had been in the twenties. He resolved, however, before this was discovered and the institute had to close down, to get the maximum advantage out of it in the shortest time. In this he was helped by the principle of secrecy. While physicists in top-secret laboratories struggled with the problem of directing a laser beam into the upper atmosphere and keeping it concentrated so as to destroy enemy missiles, two other laboratories were given the simpler task of composing a doctoral thesis for the director. This, of course, was not stated in so many words, but the scientists were assigned a subject comprising an experimental and a theoretical part, a bibliography, and everything else that a thesis should contain.

The two laboratories, at full strength, were set exactly the same assignment, but only the director knew this, as the scientists, although working under the same roof, were separated by security regulations as if by a concrete wall.[2] If this had taken place in Germany, the truth would perhaps never have leaked out; but Russian society is saved from the shortcomings of its laws by the shortcomings of its citizens. Life is only tolerable in Russia because no one obeys the regulations properly. In short, a girl who was a translator in Laboratory A, in breach of the rule of secrecy, asked a colleague in Laboratory B when she expected to finish with an American journal that the colleague had borrowed from the library. The girls got talking, and soon discovered that for months they had been translating the same articles out of the same journals. This gave them the happy idea of sharing the work by translating the articles alternately, so that each would have more leisure time. After some weeks the girls also discovered that the American technical journals kept in great secrecy at the laboratory were also in the local public library; moreover, the articles they were toiling over had been translated long ago and could be had in Russian for a few kopecks. Naturally the girls took advantage of this piece of luck, too. In this way two assistants of humble rank, with little interest in physics or politics, discovered the portentous secret of their institution, which consisted solely of the fact that dozens of scientists, highly paid by the state, were toiling to enrich a private boss, namely the director.

As might have been expected, the matter ended happily for all concerned. The top-secret establishment broke up without having discovered any super weapon, and its security rating was reduced; instead of being known only by a three-figure number, it received the innocent title of Research Institute for Impulse Physics. The translators were transferred to other institutes where they had less contact with deadly secrets. As for the director, he too is now in another institute. He is now a doctor, having duly presented his thesis—in complete secrecy, of course, so that none of its real authors could call him to account in any way. Truly a satisfactory conclusion!

Russian scientists discovered long ago that the system of state secrecy can be put to profitable use for private ends. I myself as a young journalist witnessed an example of this with nationwide, not to say international, repercussions.

In 1949 there appeared in the Moscow bookstores a work entitled *On the Nature of Viruses and Microbes*. Although it was a purely scientific work by an unknown author, it aroused great interest among the reading public. It was discussed in technical institutes, editorial offices, and social gatherings, and became the best seller of the year. The reason for the excitement was that the author advanced a revolutionary theory of microbiology, contrary to everything that had been assumed since the time of Louis Pasteur. He claimed that microbes break up into viruses and viruses can generate microbes, and that it was impossible to kill microorganisms by ordinary means—they could exist perfectly well in boiling water, for instance. Hence it was nonsense to talk about sterilization. The wretched microbes could even turn into crystals and back again into living creatures. All these strange and alarming theories were put forward by Georgi Mnatsakanovich Boshyan, an Armenian veterinary surgeon from Leninakan.

At that time I was one of the first journalists to interview Boshyan, who was established at the Veterinary Institute near Moscow. I was received by a thickset, balding man with a coarse face pitted by smallpox. Thumping the table with his fists, he told me in bad Russian that Karl Marx had discovered "new laws of social development" and he, Boshyan, had discovered "new laws of biology." He spoke with the utmost fervor and, to tell the truth, I was impressed, as were many others. Journalists and writers paid respectful heed to his theories, and a well-known dramatist, Nikolai Pogodin, wrote a play about him called *Broken Lances,* which ran in Moscow theaters for a year. Pogodin's play, however, bore little relation to Boshyan's actual career, which was much more remarkable than anything the playwright could have invented.

Boshyan's reputation was wholly the work of the director of the Veterinary Institute, a certain Professor Leonov. (Boshyan himself became a professor later on.) Leonov was in charge of this supernumerary, little-known institute when along came Boshyan with his amazing new ideas, and he realized that the crazy veterinarian could be used to great advantage at that particular juncture. The cold war was in full swing, and the intellectual slogan of the time was "Russian priority" in every field. All discoveries and inventions of any importance had been made by Russians and copied or stolen by foreigners. Countless books, articles, and lectures were devoted to proving this, and films were made about great Russian scientists of the past. Riding this wave of propaganda, Leonov told A. I. Benediktov, Minister of Agriculture of the USSR, and E. I. Smirnov, Minister of Health, that Boshyan had made a great discovery which left the Americans and British far behind. Given a chance, he would do the same again, and it was vital to support him and give him more money for experiments.

The higher-ups were convinced, and Boshyan was awarded the degree of doctor of biology without even having to present a thesis. Within a few weeks ten thousand copies of his book were printed and a large laboratory was placed at his disposal. But all this would have come to nothing if Leonov had not exploited another fetish of the period, that of universal secrecy. He insisted that Boshyan's experiments should be classified—his book contained only general conclusions—and one fine day a security post was set up at the end of the corridor leading to Boshyan's laboratory, to which there had always been perfectly free access. The director's doctoral thesis, which he wrote and presented in record time, was also given a security classification.

The meaning of all this became clear to me some years later, in the mid-fifties, when I heard a lecture by V. D. Timakov (later president of the Academy of Medical Sciences of the USSR) at a conference on the variability of microbes. Timakov said that for three years he and an academic commission composed of the leading microbiologists in the country had endeavored to check Boshyan's experiments, but had not been allowed inside his laboratory on grounds of secrecy. Meanwhile the press was trumpeting the new victory of Soviet science.[3] (This was also the period of the equally mythical successes of O. B. Lepeshinskaya, who discovered a "living substance," and of Lysenko's triumphant "agrobiology.") At long last Timakov and his commission were allowed, in conditions of absolute secrecy, to look into Boshyan's microscope, where they found that the preparations of the great transformer of microbiology consisted of nothing but specks of dirt. Dr. Boshyan, it turned out, did not have even the basic notions of microbiological technique that are required of every medical student.

The authoritative conclusions of Timakov's commission left Leonov and Boshyan without a leg to stand on; but they continued to fight, and even went on the offensive from time to time. In the end Boshyan was stripped of his degree and expelled from the

Institute, but the armor of secrecy still stood him in good stead: not a word about the fiasco appeared in the Soviet press.[4] The same procedure still works today, a quarter of a century later, when hundreds of secret dissertations, protected from scrutiny by the scientific world, make their way from the secret academic councils of secret institutes to secret departments of the Higher Examinations Board, where, being secret, they are rubber-stamped without question.

Thus, from being an element of state policy, scientific secrecy becomes an instrument of personal gain. The official concerned does not need to steal or appropriate someone else's discovery. If it is on the secret list, the laboratory chief or institute director takes the credit when reporting it to his superiors. The work was done in *his* laboratory under *his* direction, and therefore he, the official, rightly claims the praise and the reward. Declassification is unpopular with officials, in the first place because it deprives the institute administrator of kudos, but it is disliked by more senior authorities as well. In Chapter 3 I told the story of the two young scientists in V. V. Kovanov's laboratory who made the valuable discovery of ischemic toxin. My article on this subject appeared in *Pravda* on February 1, 1972. Next day Minister of Health Petrovsky telephoned Kovanov to express his disapproval: if the discovery was as important as the papers said, it ought to have been classified so as to avoid unnecessary excitement.

The minister wanted to keep secret a discovery that made it possible to regraft severed hands and feet, or to save people with heart disease. How could the chief guardian of the nation's health adopt this attitude? The answer is very simple. However highly placed an official is, he must constantly think of strengthening his position and proving his value to those at the very top. A minister is no exception, and the way for a Minister of Health to ingratiate himself is to see that the Kremlin elite are well looked after, when they need medical aid, by the special "Fourth Department" of his ministry.[5] If the discovery of Oksman and Dalin had remained secret, Petrovsky could have used it as he saw fit, restricting its benefit to the party and government higher-ups, who would have been correspondingly grateful. The declassification of ischemic toxin was in his eyes an act of theft, an attempt to deprive him of his own property.

For these reasons the tendency of the administrators of Soviet science is not to declassify, but to classify more and more strictly. The urge toward concealment is all the easier to satisfy because the laws governing secrecy, like all Soviet laws, are couched in very vague terms. This is convenient for administrators, who can twist the rules to suit their purposes, and it also means that junior officials have to play it safe. Since there is no knowing what may have to be classified in the end, a veil of political and military secrecy is cast over such things as British maritime charts, foreign scientific and technical magazines, all information concerning China or the Western Communist parties, etc., etc. In this atmosphere of security-run-mad, the Soviet "scientific" bureaucrat feels thoroughly at home.

On one occasion three institutions—a department of Leningrad University, the All-Union Research Institute for the Fishing Industry and Oceanography in Moscow (VNIRO), and the Fishery Research Institute of the city of Kaliningrad—were engaged in joint research on the use of cosmic satellites as an aid to the fishing industry. Suddenly the Kaliningrad team suggested to the university that the work be placed on the secret list. The university group was puzzled, as the subject had already been discussed in the Soviet press and was the subject of many articles in American journals devoted to physics and ichthyology. After some inquiry it proved that the suggestion really came from S. I. Potapchuk, the head of the Laboratory of Cosmic Oceanography in Moscow, who was responsible under VNIRO for seeing that the latest scientific developments were applied to actual production. Not a single hundredweight of fish had yet been caught in the USSR with the aid of satellites, and Potapchuk wanted the investigation classified so as to hush up the fact that his laboratory had been inactive and thus avoid supervision and criticism. By all accounts, he was quite successful in achieving this aim.

A Moscow mathematician, Professor Vasily Vasilyevich Nalimov, is credited with a much repeated saying: "Secrecy in Soviet conditions is the one way of concealing our scientists' shortcomings." This is true, but unfortunately it does not go far enough. The veil of secrecy enshrouding science is only a small part of the fog which envelops the whole country and every aspect of Russian life. For decades we have lived without knowing that the five-year plans were not being fulfilled, or that some of the space flights on which huge amounts are spent end in disaster (e.g., that of *Soyuz 10*). People are not told how large the nation's deficit in foodstuffs is, or how much of the food sold is adulterated. The Soviet citizen does not know what percentage of the product of his labor goes to the state; he is not told about discussions with foreign statesmen, or the incidence of epidemics in the country,[6] or the degree of radioactivity in different regions. Foreign and domestic trade figures are kept secret; so is the real value of the ruble, and the extent of known mineral resources. Even the city plan of Moscow is a secret: all the public can obtain is a rough map, not to scale. The list of secrets could go on indefinitely. It includes the size of the mesh used by Soviet trawlers on the high seas, and disasters such as air crashes, tidal waves, volcanic eruptions, and earthquakes. Science is only a small section of the vast area of secrecy thanks to which Soviet officials are able to govern without difficulty the worst informed society in the world.

There are, of course, people who do not like this system, especially among writers and scientists. There are even some who demand publicity in the words of Mendeleyev, who once said: "Science, by its nature, cannot be secret: it must be public or it is not science."[7]

A candidate of economic sciences working in a research institute once told me that he had been trying for six months to obtain access to the Central Statistical Office in order to analyze production figures. When he finally got permission and presented himself at the Office, he was made to sign an oath that he would not quote any of the figures in his articles and monographs—the whole body of information was top secret.

This, of course, is infuriating to an economist; but after all, how can the authorities be expected to divulge the true figures relating to the five-year plans, when for many years they have been only 60 or 70 percent fulfilled, while the Statistical Office proudly announces every six months that the degree of fulfillment is 100 percent or more? It is easy to imagine the chaos and argument that would ensue if economists were allowed to give their own version of the facts; but secrecy, thank heaven, makes it possible for what is published to coincide exactly with what is desired.

Again, the standard of living in the USSR is a dark secret, with the excellent result that no one gets riled or complains to the government about it. True, there are 300,000 families in the Soviet Union which, under instructions, regularly send precise details of their income and expenditures to the Central Statistical Office. Until 1968 these figures, although not published, were at least collated, and economists were allowed, in great secrecy, to consult them. But for some years now they have been top secret, and analyses are no longer available.

Let us turn to the domain of space flight. Much is written about space flight as a means of scientific investigation, but experts have told me that of the huge sums invested in the space program, science accounts for at most 1 percent; the balance is for defense. In the same way, annual figures of increased state investment in research must be taken with a grain of salt. The sums are indeed increasing, but the lion's share is accounted for by research institutes that do not even deserve to be called scientific. Here again, secrecy is all.

Infringements of state security are treated in different ways according to the purposes for which they are committed. If the offender's object is merely to grab a share of profit for himself, his superiors will not take it too much to heart: they may shove him away from the trough, but rather for form's sake and without undue violence. Boshyan, for instance, is still employed in a Moscow research institute and is said to be flourishing, and he is not the most remarkable case. Not long ago N. G. Basov, an academician and winner of Lenin and Nobel prizes, was caught passing off another's discovery as his own. He was lecturing on physics to an academic gathering when his colleague, Yu. B. Khariton, openly warned that if Basov ever again claimed, orally or in writing, to be the originator of the ideas he was putting forward, he (Khariton) would demand the publication of articles by A. D. Sakharov going back to the fifties. This silenced Basov, as his plagiarism was quite obvious to specialists. However, if it had come to a showdown on the point of declassifying Sakharov's articles, Basov would have won: it would have been a *political* issue, and no one would have allowed them to be published before the statutory twenty years had elapsed.

But, while infringements of the system for one's personal profit are looked on with understanding if not actual sympathy, it is a different matter if anyone seeks to tear down the veil of secrecy for disinterested reasons, for the sake of the public good. Then the authorities really open their eyes, and woe to the offender! A recent incident of this sort made me refresh my memory, from letters and diaries, on the case of Yulo Vooglayd, an Estonian candidate of philosophical studies. Here are some extracts.

Diary. Moscow, June 6, 1973.

Sad news from Estonia. The party Central Committee in Tallinn has on its agenda today "The Organization of Sociological Studies." The matter at issue is that the authorities are displeased with Yulo Vooglayd, a talented young candidate of sociological studies from Tartu University, for setting up a laboratory whose work they regard as irregular. The proposed solution is already known: it is not to close down the laboratory, although they could, but to frighten enterprises which have concluded working agreements with it. The sociologists get 95 percent of their funds from contracts with plant managers whom they advise on production methods. If these people go back on their contracts, Vooglayd will have to close down the laboratory. That is what the Central Committee wants, as apparently the sociologists "know too much."

Diary. Tõrva,[8] Estonia, July 19, 1974.

Yulo Vooglayd, the sociologist, came here from Tartu. A quiet, slow-mannered chap, the Viking type—bearded, with blue-gray eyes. He is one of the few Estonian intellectuals I have met who isn't always whining about the nationality problem but gets on with his job—he works hard and does a great deal. He has set up a big laboratory at Tartu University. Other sociologists say, half-jokingly, that he knows *everything* about Estonia. That is not absolutely true, but out of ten questions of mine about Estonian ways and culture, family relationships, food, housing, attitudes to work, and so on, he produced the answers to eight from his filing cabinet.

He believes that his studies will provide insight into the workings of society, not only in Estonia but in the whole Soviet Union. He outlined a scheme showing that his laboratory was on the way to creating a model of the social life of the whole country. Who knows, maybe the pattern derived from a small society like Estonia really will enable people like Yulo to plot the development of our whole nation. . . . He has four children and a dog, his income is very small but he shows no sign of worrying about his career. He hasn't even bothered to present a doctoral thesis—he could do so very easily, but he's too busy!

The sociological laboratory at Tartu is constantly in trouble. The

last complaint came from Rutkevich, the director of the Institute of
Practical Sociological Studies in Moscow; it was discussed in Moscow first and then by the Central Committee in Tallinn. Rutkevich
said the laboratory was full of nationalists and Zionists—there are
some Jews on the staff—and ought to be closed down. . . .

Yulo is fond of sport and is the Estonian yachting champion. He
has a simple and dignified manner. He invited me to go yachting
with him on Sunday. He plans to spend twelve hours sailing first to
Saaremaa Island, where he was born, and then to Riga. Looking at
him, it struck me that in Moscow you seldom see this combination
of vigorous mental activity and physical health—it is more of a European characteristic.

Diary. Tõrva, August 6, 1974.

Yulo Vooglayd invited me to Tartu, where I gave a talk to his
staff on "why a scientist needs a conscience." They listened attentively, and I saw many good faces among them. Vooglayd remarked
that morality was specially important to a sociologist.

Diary. Tõrva–Tartu, August 15–16, 1974.

Spent two days with Yulo Vooglayd. We talked a lot about Estonia and the Estonians. Yulo had some marvelous stories to tell. He
also unexpectedly showed us a new side of himself, that of a family
man. He spent a whole day running around town in search of a
metal tube for his thirteen-year-old son, who is competing in a
yachting race the day after tomorrow—the tube is needed to
strengthen one of the sails. Yulo went round all the shops, garages,
factories, and sporting clubs in town, and finally got what he
wanted. Coffee with the Vooglayds—almost the whole family present, except the smallest daughter and the dog. Kay and Katya, aged
fourteen and sixteen, are fresh-faced, suntanned, sport-loving girls.
They don't talk much, but are perfectly simple and unaffected. Kay
is already a noted athlete. Katya is studying several Slav languages
and intends to go in for linguistics. But we liked Tomas best of all:
flaxen hair, a clear, manly gaze, not talkative, but full of boyish dignity—the epitome of what a lad of thirteen should be.

Letter from N., sociologist. Leningrad, December 1974.

. . . It would of course be better for Vooglayd to explain it himself, but, since you ask me, I will try and describe the purpose of his

work. Unfortunately you aren't familiar with sociological terms, so I will use everyday language. Vooglayd's laboratory is devoted to studying the Estonian way of life: food, housing, consumer goods, industrial and family relations. . . . The inhabitants of the Republic are divided into eight types according to life-styles and material circumstances. Thanks to the work at Tartu we now know what kind of houses, furniture, cars, refrigerators, and other goods the population wants, and which it doesn't. We also know about family relations: how much each group wants or doesn't want to have children, what their customs are as regards marriage and divorce. The laboratory also studies industrial conditions. Having studied twenty-eight motor transport establishments and several factories, Vooglayd has identified the factors which prevent their working efficiently. His staff has shown experimentally that present-day industrial workers could raise their output by 150–200 percent, and they have indicated practical ways of achieving this. Having studied the "Estonian way of life" they have also produced concrete proposals for improving trade and transport in Estonian cities. Vooglayd is also studying the effect of the mass media—newspapers, radio, television—on the Estonian public. In addition he has developed a theory of communication that is as yet unknown in the West. He believes that in our society it is possible to increase greatly the effect of education and propaganda on individuals and entire groups, by placing them at the point of intersection of several channels of communication. . . . As far as I know, however, he preferred not to publish this scheme, and he is said to have burned the manuscript containing it. . . .

Vooglayd's studies and calculations already make it possible to formulate useful economic and political recommendations. Labor efficiency can be stepped up, and the production of many goods can be planned so as to prevent overstock; sociologists can point out to information and propaganda services where they have gone wrong, and can aid demographers to predict family trends. But apparently the Estonian leaders aren't interested in all this valuable information.

Diary. Tõrva, August 14, 1975.

It is just a year since we last saw Yulo and his wife, and here we are together again at Tõrva. There is a striking change in Yulo: last year he was irresistibly confident, but now one feels there has been a kind of crack-up. Although he is still in his forties and hale and hearty, the last few months have had a crushing effect on him. On

July 1, without a word of explanation, his laboratory was closed by the rector's order. A hundred of his pupils and staff staged a midnight procession in front of the university building, with torches and farewell songs. So, after nearly ten years of existence, the laboratory is no more. Yulo thinks the Tartu and Tallinn authorities made the decision a long time ago: there were special resolutions by the Central Committee and warnings by the Tartu bosses, and the rector simply bowed to the inevitable.

The work of Yulo and his team showed up the incompetence of factory and commercial management, the indifference of workers, and the lack of responsibility at all levels of the social structure. Their facts and figures testified to the economic and spiritual decline of the Republic. I said to him: "But you hardly published anything —how did they know what your conclusions were?" "That's true," he said, "if we had published sooner, no doubt they would have closed us down years ago." He partly blames himself for what has happened. Last December he was asked to give a talk to the Republican planning commission, and in it he drew attention to several defects of the economic system. His audience said nothing, but he had a feeling that he would not be forgiven. They could not put up with the fact that an outsider knew the truth about *their* economy, *their* power, *their* system of administration.

After the laboratory was closed down, the rector proposed that Vooglayd should be expelled from the party. This means that he will not get another job in any department of sociology. . . .

Letter from M. A. Popovsky, member of the Writers' Union of the USSR, to I. G. Kebin, First Secretary of the Central Committee of the Estonian Communist Party, dated September 10, 1975.

Dear Sir,
    . . . The Estonian sociologist Yulo Vooglayd is threatened with complete exclusion from scientific work. His laboratory has been closed down, and this gifted scientist in the prime of life has been expelled from the party. Apparently his expulsion from Tartu University is also being considered. . . . I can say with complete confidence that the laboratory of which Vooglayd has been in charge for nearly ten years was one of the most impressive scientific establishments known to me, in its wealth of original ideas and the friendly atmosphere of quiet efficiency which prevailed there. It dealt, moreover, with the most vital questions affecting the life of the Republic. . . . The value of the laboratory's work consisted

above all, it seems to me, in providing managers of enterprises with full information on areas of concern to them. By collecting and analyzing the replies to thousands of questionnaires the sociologists organized a feedback from society to management. . . .

I know Vooglayd as a man of transparent honesty with a passionate devotion to science. Loved and respected by his staff and pupils, he is a good lecturer and a true researcher. I shall continue to work on my book about Vooglayd and his team, because I am certain that their work is the kind of example that should be set before young people. I also feel sure that even before I finish the last page, the Estonian party leaders will have restored the good name of my literary hero. . . . I say nothing of what this means for sociology, but I send you this letter in the hope that you will not be indifferent to the fate of a human being.

Reply to the foregoing, undated.

Dear Sir,

. . . [your] fears concerning the development of sociology in the Republic are unfounded. On the contrary, work is proceeding with a view to improving the quality and efficiency of sociological studies. This is the purpose of the recent reorganization in Tartu University, which in no way signifies the liquidation of such studies.

As regards Comrade Vooglayd, it must be noted that unfortunately he committed various mistakes and improper acts of a political character. His case was considered and decided upon in full conformity with the Rules of the Communist Party of the Soviet Union. We are prepared, if necessary, to help you obtain fuller information on the questions you raise.

[Signed]: E. Grechkina, head of the department of science and education of the Central Committee of the Estonian CP, and O. Utt, head of the department of culture.

Letter from I. G. Kebin to M. A. Popovsky, dated November 13, 1975.

On receipt of your first letter I instructed the officials of the Central Committee of the Estonian CP who are most conversant with the case of Y. Vooglayd to send you a reply. I fully agree with its contents. It seems to me that they have made a very helpful suggestion in offering, if the case really interests you, to provide fuller information in an interview at the office of the Central Committee.

Letter to M. A. Popovsky from V. Renzer, assistant to First Secretary of the Central Committee of the Estonian CP, dated January 15, 1976.

In answer to your telegram I am instructed by Comrade I. G. Kebin to state that a reply on the matter you raised was given by Comrades Grechkina and Utt, heads of the competent departments of the Central Committee of the Estonian CP, in their letter of October 21, 1975. This reply was confirmed in Comrade Kebin's letter of November 13, 1975. Fuller information on the subject can be provided *only* in an interview at the office of the Central Committee of the Estonian CP.

I did not take up the invitation to discuss the matter at the Central Committee office, because I knew that Vooglayd's fate and that of his laboratory had been irrevocably decided. He is no longer a member of the university, and is no longer at Tartu; he can no longer teach pupils or pursue an academic career. Those who destroyed his laboratory and killed a living scientific organism did not even dare to state publicly what they had done—it was, of course, another official secret. I did not go to Tallinn, because I knew the version I would be given there. The party authorities tried for a long time to trump up a charge that would bring shame and discredit upon Vooglayd. They did not succeed, because everyone in town knew and loved him. Then the KGB took a hand: he was called in seven times for interrogations lasting five or six hours, with the object of making him confess that he had drafted the manifesto of a mythical Estonian National Council. As the KGB well knew, he had done no such thing: the son of a peasant family and its first intellectual member, his only interest throughout life had been science.

The political charge finally had to be dropped; but some way had to be found of discrediting him, as he knew too much. A special commission of the Central Committee was set up to examine his "case"; it proceeded to Tartu, questioned dozens of people, and finally obtained what it wanted.

Some years before, a young assistant at the laboratory had been given 1,500 rubles by his peasant father. He decided to use the money to buy a dilapidated farmstead near Tartu. Aided by friends and colleagues, he restored the house, planted a garden around it, and turned it into a club to which the young sociologists came on weekends to sit around the fire and talk about art and science, take a sauna bath, or dig potatoes. Artist friends provided pictures to decorate the walls, and an album was kept on a table in which members and visitors could record their thoughts, activities, and aspirations. A young poet who was there as a guest might contribute a few lines, or a young journalist might write a short sketch. This impromptu "journal" was kept in two copies; but neither the albums nor the paintings were submitted to the censor, and this illicit proceeding became one of the principal charges against Yulo Vooglayd. "How could you allow such a thing?" the commission demanded angrily. There was, however, an even more serious offense. On a stone slab in the main room two army helmets were discovered, one of German and one of Soviet type: the young people used them to bake potatoes in, a purpose to which they are excellently suited. Quantities of such helmets are still to be found in the forests of Estonia, but the commission cast a sinister light on the affair; they also pounced on the fact that the Soviet helmet had bullet holes in it while the Nazi one did not.

That was all. A battered helmet, shot through in a war which was now part of history, served as a damning piece of evidence against Yulo Vooglayd and his laboratory. An appeal board of the Central Committee of the CPSU, presided over by Pelshe, a member of the Politburo, confirmed the sentence passed in Tallinn: Vooglayd's laboratory was closed, and he himself was expelled from the party and from academic life. His real crime was to have obtained an insight into the social mechanism of Soviet life. For secrecy is the law of the land, a law overriding all others.

CHAPTER 5

WE AND THEY, OR "RUSSIAN TIME"

"Other countries are different, but we live in an ordered community! I traveled all over Europe for six months and no one stopped me, but it couldn't happen here! You can't enter or leave the country without permission: every one of us is suspect—after all, he might be a malefactor!"*

The opposition between Soviet science and world science is a matter of constant variation: sometimes there is a thaw, sometimes a chill, sometimes a severe frost. What there has never been within the memory of my generation is an ordinary equable climate: for the scientific "weather" at any given moment is determined not in laboratories and institutes, but in the offices of party bureaucrats, and it depends exclusively on the political atmosphere prevailing at the time. In the last years of Stalin's rule, when the cold war was at its height, Russian scientific chauvinism surpassed itself in absurdity. From his newspaper or radio the citizen learned, for instance, that cybernetics represented the darkest and most despicable aims of world imperialism, and that "formalist" geneticists had only thought up genes and chromosomes because, lacking the vital clue of dialectical materialism, they were unable to comprehend the world in all its complexity. Other equally instructive facts were to be learned about mathematical linguistics, philology, and bourgeois anti-Pavlovian physiology. In a journal dating from the late forties, which I keep with the same reverence an archaeologist might bestow on the fragments of a Greek vase, one may read that "the law of the conservation of energy was discovered by the Russian scholar M. V. Lomonosov, not the English brewer Joule or

* M. E. Saltykov-Shchedrin, *Complete Works*, Vol. 11, p. 487.

the German physician Helmholtz." Those were the days! Not only did the learned writer refer to Joule as a brewer (it was his father who had owned a brewery, and Joule of course was an eminent physicist), but it was simply impossible for the Soviet reader to find out that the law of the conservation of energy was in fact discovered not by Lomonosov but by Joule, Helmholtz, and another physician, Julius von Mayer.

Scientific chauvinism and nationalism are not confined to a single age or country. The British scientific community was incensed when Sir Humphry Davy accepted a prize from Bonaparte at the height of the Napoleonic Wars. During the Franco-Prussian War, Louis Pasteur angrily threw away papers sent him by the German scientist Robert Koch. Coming to more recent times, I read with astonishment in the *Memoirs* of the French surgeon René Leriche (1879–1955) how, while insisting on "truth, scientific and historical truth," the author wrote indignantly and, alas, unjustly that the credit for pioneer work in operative asepsis had been given to the German Bergmann whereas it belonged to the Frenchman Octave Terrillon. Scientific rivalry and political passions will no doubt cause similar misunderstandings in the future. But it is hard for a Western scholar to imagine to what a pitch these human weaknesses are developed by half a century of propaganda, when all the efforts of the state bureaucracy are consciously devoted to inflaming them.

Chauvinism of this sort afflicts the law-abiding scientist with schizophrenia. If he wants to be a loyal citizen he must at all costs extol everything Russian and disparage the work of foreign countries. When he meets a foreign colleague he must give the impression of being a freeman, but he is strictly forbidden to invite the foreigner to his home, nor can he write frankly to anyone in a foreign country. When he comes to his office after visiting an exhibition of American machinery or domestic appliances, he is careful to keep his impressions to himself for fear of annoying his superiors. He remembers only too well that when he appended too long a list of foreign authorities to his doctoral dissertation, it was mercilessly cut down and was considered a black mark against him.

We may be told that these excesses are a thing of the past, and that in our day, when *Soyuz* and *Apollo* spaceships engage in joint operations, it no longer makes any sense to talk of opposi-

tion between Soviet and Western science. But is this so? Certainly
Soviet scientific policy has altered since the détente set in, as was
confirmed to me not long ago by Aleksandr Aleksandrovich
Marinov, editor in chief of the Znanie publishing concern[1] and a
former colonel in the Political Department of the Soviet Army.
Our conversation was not a brief or a casual one. I had a contract
with Znanie for the publication of *Why a Scientist Needs a Conscience*. The manuscript had been in their possession for over two
years, but the editor in chief, the scientific editor, and the regular
editor kept pressing me to make changes on political grounds. One
day I walked into Marinov's office and saw with annoyance that
another fourteen pages had been deleted. The text, so precious to
me, was scored all over with blue-pencil marks like bruises. Particular offense had been given by a quotation of mine from a document of 1920 in which a group of Soviet physicists had appealed
to the authorities to allow freer contacts with foreign scientists:

> The isolation of any country condemns it to backwardness and
> stagnation in the field of science. . . . Science is international by its
> very nature. It represents the result of the collective experience of
> all humanity, and in order to develop without interruption it re-
> quires uninterrupted contacts and interchange between human
> beings and in particular the scientists of all countries, so that they
> can at once make use of one another's discoveries.

To this I added on my own account that experience had more
than once shown the necessity and inevitability of free contact
among scientists, the free interchange of ideas, the sharing of discoveries and achievements.

This last remark had been blue-penciled with special violence,
and in the margin was written: "No limitations at all? No political
or military secrets? Absolute freedom??" The question marks bris-
tled sarcastically. I sat there dejected—it was the fifth interview at
which pieces of my text had been torn out, like flesh from a living
body. The editor in chief clearly felt sorry for his victim, and he
launched into a speech with the evident object of helping me to understand the wisdom of state policy.

"The mistake you are making," he explained in an earnest, confidential tone, "is that you take too seriously the slogans of interna-

tional détente. Of course we say to them, 'Come on, let's have exchanges of scholars and ideas.' But it isn't all as simple as that. After all, we have our own interests, we're a socialist state, and we can't forget that capitalism is a predatory system." Then, becoming even more confidential, Marinov leaned across the table and lowered his voice to a whisper (although we were alone in the room) as he added:

"The Central Committee held a conference for propagandists and told us straight out: the whole purpose of scientific contacts with the West is to get as much from them and give them as little as we can. Of course we need some Western discoveries and technology, but we have no intention of cooperating fair and square with them. Do you understand? No rose-colored spectacles, they told us; don't let the wool be pulled over your eyes. And here you go writing about the necessity and inevitability of free contacts."

Colonel Marinov looked at me with sorrowful reproach, as a schoolmaster looks at a pupil to whom he has given a low mark for the nth time—my political naïveté obviously grieved him deeply.[2]

Well, a senior army commissar and a party propagandist of thirty years' standing ought to know what he is talking about; and in the books to which he gives his imprimatur it is made quite clear to the reader that détente is all very well, but there must be no kind feelings toward Western scientists, since they are merely servants of the ruling capitalist class. Of course there are progressive-minded scientists in Western countries too; but "are they not left in freedom simply in order that the monopolies may somehow exploit their scientific achievements?"[3]

Such is the general line of Soviet scientific policy in the period of détente. We are not speaking of the ill will of some dull-witted institute director, laboratory chief, or editor of a scientific publishing house—this is traditional state policy, deeply rooted in the consciousness of the huge army of Soviet scientists. Translated into everyday language it means: "Foreigners are not to be trusted; a foreigner is always a secret enemy; international cooperation, *Apollo-Soyuz,* the Pugwash Conferences, leading articles in newspapers—all that is just official eyewash. It's all right at the top level, but down here among the 'million' we have to be doubly careful of contacts with foreigners and always remember how dangerous they can be."

Official suspicion and unfriendliness toward everything Western obliges Soviet scientists to behave with fantastic ingratitude. In May 1974, during a visit to the agricultural research institute at Krasnodar, I was sitting in the office of my old friend Mikhail Ivanovich Khadzhinov, an academician and the subject of several of my books. A letter arrived from Professor Lambert of the University of Illinois, head of the international bureau of geneticists concerned with developing varieties of corn, informing Khadzhinov that he had sent to Krasnodar a large assortment of seeds with characteristics of special interest to his Soviet colleagues. This letter (dated April 29) put Khadzhinov in a gloomy mood, as he described to me how badly we behave toward our foreign contacts. Plant breeders in the United States, France, and Britain respond to every request on the part of Soviet scientists, but for the latter to send material abroad is a problem of the first order. Many departments have to be consulted; if the would-be sender is not persistent enough, his request will get stuck in one or another of them; but if he insists too much, the party officials will get suspicious and withhold their consent for political reasons.

On one occasion the Americans asked the agricultural institute at Kishenev to send them specimens of a strain of corn developed by the Moldavian breeder Miku, in which necrosis of the tassel was transmitted genetically; but the officials in charge of scientific matters would not allow the seeds to be sent to "imperialists." This was at a time when the whole Soviet experiment with hybrid corn, directed toward increasing the protein content, was based on material provided by American geneticists! To Khadzhinov's credit I must say that he himself is strongly in favor of promoting such exchanges with colleagues all over the world, including Illinois and Ohio. Thanks to his personal authority he is able to overcome many barriers and prohibitions imposed by the Krasnodar authorities; but even his contacts with foreign scientists are constantly interfered with. Every time there is a question of his pupils going on a course in the United States or to an international geneticists' congress, the formalities drag on through many weary months.

Fear and malevolent propaganda lead to truly incredible situations in research institutes. The world's principal biochemists, including several Nobel prize winners, will remember the seventh International Biochemical Congress held at Riga in summer 1970. More than two hundred papers were read, and the Congress is said to have made a good impression on the foreign visitors. Its host, as they may also remember, was Solomon Aaronovich Giller (1915–1974), director of the Institute of Organic Synthesis of the Latvian Academy of Sciences. I knew Giller for many years and, to tell the truth, our relations were somewhat mixed. I wrote about him several times,[4] but was far from sharing his ethical outlook in all respects; he was, however, undoubtedly a gifted scientist and a good organizer. He was also one of the two or three Soviet scientists whose medicinal preparations were sold on the world market. This is noteworthy if one remembers that the Soviet drug industry is still in an embryonic stage, that the public has suffered from shortages in this area for decades, and that the most essential preparations are imported. Giller, however, managed—not on his own, of course, but through the state trading organization—to market one of his preparations in Japan and another in Sweden. The Japanese paid $500,000 for an anticancer preparation made in Riga, and the Swedes $7 million for a license to produce mitandion, a drug to cure epilepsy. Giller returned from his last trip abroad in a jubilant frame of mind, and reported accordingly to the Presidium of the Latvian Academy. However, instead of gratitude he received a sharp admonition from the vice-president: "Who authorized Academician Giller to give our scientific secrets to the capitalists?" The president then launched into a long diatribe, mentioning among other things that he himself had been wounded at the front in World War II and could not get the penicillin he needed, because the British would not divulge the secret of its structure and manufacture. And now here was Giller going to the capitalists and "giving away" (for only $500,000) the precious achievements of Soviet science.

There is no difficulty in refuting the myth that the British refused to reveal the secret of penicillin. But it may be more interesting to point out that eighteen months earlier the president of the Latvian Academy had welcomed the biochemists attending the International Congress, with a speech full of references to friendship, cooperation, and the unity of scientists throughout the world. Very likely this speech was reported in foreign papers and made an excellent impression. But only a year and a half later, while détente was still officially in full swing, the same president made a speech which essentially opposed friendship, cooperation, and mutual confidence. Of course, the audience was different: the Presidium is home ground, and there you can say all you think, or all you are supposed to think.

Did Giller reply by reminding his colleagues of the duty of internationalism in science, or of the fact that Fleming, Florey, and Chain abstained from patenting their invention so that it could reach Soviet frontline hospitals more quickly? No, he did not: "I was badly frightened," he confessed to me afterward, "and started making excuses." As it happened, the excuses were unnecessary: the Latvian deputy premier, who was at the meeting, explained to the academicians that "the creation of scientific values which can be turned to account on the world market is not to be regarded as a crime but rather as a patriotic duty. There is no ban on trading such discoveries through the appropriate state machinery." Giller was saved; but what are we to say of the attitude of his academic colleagues?

To be fair, they could not have been expected to behave otherwise. At every briefing by the party's Central Committee and Regional Committee they had been told again and again that contacts with foreigners were dangerous—all the enemy wanted to do, either in Russia or through meeting Soviet citizens abroad, was to carry on propaganda and espionage. And should not scientists take warning from the fact that it took months, as a matter of course, to "clear" the application of one of their number to attend an international symposium? The president and vice-president might have expressed themselves in old-fashioned terms, but essentially they had reacted as law-abiding citizens should. The same may be said of Professor Polshkov, the director of a geophysical institute under the Academy of Sciences of the USSR, who recently made an even more curious speech to the Academic Council of his institute. But let me relate the facts in order.

Vadim Minukin, a candidate of physical sciences and an employee of the Institute, was working on a problem which was to be the subject of his doctoral thesis. Toward the end of his experiments he published an article in a certain Soviet journal. This must have described some interesting data or methods, for it provoked a letter from a laboratory in Los Angeles. The writers complimented Vadim on his investigation, which had gone further than their own, and suggested that it would be useful for the two teams to collaborate. The letter arrived at the Institute but was not delivered to Minukin, who only heard of it indirectly. At the next meeting of the Academic Council, the director announced that he had received a letter from the United States which was clearly a "provocation." The Americans were trying to divert Soviet geophysicists from the true path by proposing a completely unrealistic method of investigation. To cooperate as they suggested would hold up Soviet research or lead it into a quagmire. The best thing to do was to abandon the line of inquiry in which the Americans were interested, and erase it once and for all from the Institute's plan of work.

The Academic Council obediently endorsed the director's proposal, and next day Minukin was informed that the work he had been engaged in for years was of no interest to anyone. All his experiments and observations were to be struck from the list of projects as unreliable and scientifically worthless, and of course there could be no question of their serving as material for a doctoral thesis.

It is not only directors of institutes and prominent academicians who are frightened by the intrigues of foreign scientists: the poison of suspicion and distrust infects each and every one of the million. It is particularly strong in provincial institutes and those subordinated to ministries and not to the Academy, where the fear of contact with foreigners is intensified to panic and terror. I remember a controversy that arose at the All-Union Institute at Ramon near Voronezh, which carried out research into the genetics, biochemistry, and cultivation of sugar beet. Ramon, a small locality, might have been called the sugar beet capital of the USSR, at all events during the fifties and sixties when Avedikt Lukyanovich Mazlumov (1896–1972) was in charge. Mazlumov, a member of the Lenin Academy of Agricultural Sciences, was visited by beet specialists from both Europe and America. When working on his biography I went to Ramon several times and discussed the Institute's work with members of its staff. One of these discussions, while full of interest at the outset, ended, I will not say in a quarrel, but at any rate in an estrangement between us.

The staff of the Institute is mainly composed of agronomists who have graduated from the Agricultural Institute at nearby Voronezh. They are experienced and know their job but are not as a rule remarkable for cultural interests, being satisfied with a film and a dance to the accordion at the village club. For this reason I particularly noticed Iraida Vasilyevna Popova, a woman in her fifties who was a candidate of science and a phytopathologist. She too had spent many years in the fields of Ramon, but had not lost interest in literature, music, and social relations. Her thesis on sugar beet pests, which she had recently presented, had been commended by experts for its thoroughness and had received recognition abroad—it was indeed this circumstance, harmless in itself, which spoiled our relationship. Iraida Popova told me that, two months previously, she had had a letter from a British entomologist asking for a reprint of her latest article. But she had not replied and was not going to, as she was afraid of "relations with foreigners."

I ventured a joke about "foreign devils," but it fell flat. I then reminded her that such an eminent scientist as Nikolai Ivanovich Vavilov had attached great value to scientific internationalism. I even opened a copy of my book on Vavilov, which had come out shortly before, and read her what the famous biologist had said about the many friends all over the world who, despite officialdom, had helped him to carry out his various expeditions.[5] But it was no use; nothing I could say made the slightest impression on Popova's bedrock certainty. We did not finish the argument, as it was time for me to drive to the station and catch a train for Moscow; but Popova promised to write and explain her views at greater length. About a week later I duly received her letter, which I quote almost in full:

> I do not know whether I shall be able to overcome my "spiritual slavery," as you were pleased to call it, but my *system* [all emphasis by I. Popova] is a firm and reliable one: to serve the interests of *Soviet science*. If any acts of mine run counter to this in the very slightest degree, I cannot call myself a Soviet scientist. . . . My correspondence with scientists in the socialist camp is friendly, unconstrained, and uncensored.[6] But in the case of material sent to capitalist countries, much may be read between the lines and mis-

interpreted. We know many instances of meetings with visitors to this country whose main purpose is to look out for defects and who present everything they see in the worst light. . . . Consequently, in deciding whether to reply to the Englishman I must think not of myself but of the whole scientific community and our prestige in general. This is a matter of *cast-iron* consistency, as the phrase goes, and clearly it is not likely that your advice can help. . . .[7]

Comment would be superfluous; I can only add that I have no doubt of Popova's sincerity.

"Manipulated science"—that favorite child of the party bureaucracy—is easy to handle, submissive, and totally reliable, but it has one defect: it is devoid of creativity. This does not mean that the "manipulated scientists" are nonentities. Some of them are gifted, but the more talent they have, the more they chafe and seek to escape from the official harness. This is not a matter of vanity, but rather of the very nature of science. Restraint is not irksome to mediocrity, which has no objection to being muzzled. But when a real scientist discovers or invents something out of the ordinary and breaks into a new field, he is at once impatient to know what other scientists think about it in other countries—America, Albania, Australia, the Åland Islands, or wherever. What is the reason for the strange agitation that impels him to compare notes and views?

Really fundamental discoveries relating to the problems of life and death, the structure of matter and the universe, are never purely local in origin. Great ideas and problems hover in the air for a long time, exciting minds and arousing passions; attempts are made to solve them, with varying success, in different corners of the globe. This was the case with anesthesia, antibiotics, the nature of light, Mendel's periodic table, nuclear fission, and many other key problems. Success alternates with failure, hope with disappointment. The world of investigation is like a saturated salt solution in which crystallization is about to take place. Only a slight impulse, only an opportunity, is needed. The moment of greatest tension comes just before the crystallization of a great discovery. During those days and hours, information about what has already been achieved and what has not becomes a matter of decisive importance to the researcher. The faster information is transmitted, the sooner new ideas and discoveries will make their appearance. This is a law of science, and it explains why so many symposia, conferences, and congresses are held throughout the world. Scientists do not begrudge the time and money spent on meeting one another and discussing their problems. It has been estimated that a present-day scholar gets 70 percent of his information from direct contacts with his colleagues. The free exchange of information is a vital feature of contemporary international science, and the lack of it is what bedevils science in the Soviet Union.

The rank and file of Soviet scientists have no contact with the outside world; they alone have no part in the international dialogue. True, they are not so utterly cut off as in Stalin's day; an occasional breath of fresh air is wafted across the barriers. But the contact they enjoy is so meager that the ablest of them are condemned to vegetate and to work at half the rate. As a result, even the most brilliant ideas of Soviet scientists come to fruition too late or not at all. It is a commonplace for ideas that originate in Russia to find application only in foreign countries. The retardation of Soviet science, about which A. D. Sakharov wrote recently, is a direct consequence of the restriction of scientific intercourse.

It is in no way surprising that Iraida Popova, or the administrators of the Latvian Academy, do not notice this state of affairs and are not irked by it. Their own scientific attainments, which are not so remarkable, can for the time being derive sufficient nourishment from their own sources. But a chemist from the Science City of Chernogolovka, whose discoveries make it possible to extract nitrogen fertilizer and fuel from the air without appreciable loss of energy—such a person is part of world science, and belongs with every fiber of his brain to the network of discovery that is going on in foreign countries as well as at home. It is torture to him if he cannot meet British, French, and American colleagues, so as to hear about their achievements as soon as they take place and to show them how his own work is progressing. He suffers from being turned into a country bumpkin of science.

This situation was described to me in February 1976 by Gerz Ilyich Likhtenshtein, who was in a state of depression because he had, for the nth time, been refused permission to travel abroad. The occasion was a symposium in Italy on the kinetics of chemical processes; but Professor Likhtenshtein is also forbidden to go to

1. Academician Nikolai Vavilov (1887–1943), founder and first president of the Academy of Agricultural Sciences in Moscow. Geneticist, geographer, horticulturalist, traveler, Vavilov won the Lenin Prize in 1926. He was arrested on August 6, 1940, and died of hunger in Saratov Prison on January 26, 1943. This picture was taken in 1928.

2.   Another photograph of Nikolai Vavilov. This picture was taken two weeks before his arrest in July 1940.

3. Mikhail Hadjinoff, member of the Academy of Agricultural Sciences and disciple of Nikolai Vavilov. An outstanding geneticist, he was the originator of the best varieties and hybrids of corn in the USSR, and was called the "Hero of Socialist Labor."

4. Microbiologist, Professor Nikanorov, director of the Microbe Scientific Research Institute at Saratov. He worked on aggressive and defensive biological weapons in the secret bacteriological center at Suzdal. Arrested in 1930, he was shot in 1932 or 1933.

5. Talented microbiologist Dmitri Go lov (1898–1937). He worked in the Mi crobe Research Institute at Saratov, an was first in the USSR to isolate the tula remia bacillus. Twice arrested, he wa exiled, and then shot.

6. Solomon Giller (1915–1974), professor of chemistry, member of Latvian Academy of Sciences. Director of the Institute of Organic Synthesis in Riga, he discovered several valuable medical preparations, and was one of the few Jews to head a scientific research center in the USSR.

7. A unique figure in Soviet science, Professor-surgeon Valentin Voyno-Yasenetsky (1877–1967) combined scientific and religious activities. For forty years he was a bishop of the Russian Orthodox Church, and for the monograph "Essays on Suppurative Surgery" (1946) he was awarded the Stalin Prize. He spent twelve years in jail and exile for his religious beliefs.

8. Gevork Boshyan, a veterinarian from Armenia, made an attempt, in the early 1940s, to disprove Louis Pasteur's discoveries. He was awarded the doctor of sciences degree without defending a thesis, and his book was hailed as an achievement in Marxist biology; 100,000 copies were published. In the early 1950s, Boshyan was denounced as a charlatan.

9. Two friends: Joseph Stalin and Trofim Lysenko, academician. In the 1940s such monuments decorated the plazas and streets of many cities in the USSR.

10. One of the last encyclopedic scholars of the Soviet Union, Professor Aleksandr Lyubishchev (1890–1972). Mathematician, biologist, philosopher, author of several studies of professional ethics in Soviet science, he was constantly persecuted for his views. Most of his work is unpublished.

11.   Outstanding Soviet geneticist Professor Victor Pisarev (1882–1972). He was arrested several times for so-called "political crimes," and under the threat of death, he issued a false statement in a Soviet jail against his friend, Academician Nikolai Vavilov.

the United States, where he has a long-standing invitation from Professor McConnell of Stanford, or to accept invitations from many other parts of the world. Gerz Ilyich, a robust man with a mop of curly black hair and a lively, good-natured expression, was seriously disturbed: a few more obstacles of this sort, and he could say good-bye to his international reputation. Nowadays, if a scientist does not go to a congress it means he has nothing to say.

Who is it that prevents Likhtenshtein from going abroad? Like most Soviet citizens, he is well enough trained not to tell anyone his inmost thoughts. Perhaps in secret he curses the fact that he is Jewish, or reviles the official in the foreign department of the Academy of Sciences who persists in "losing" the file with his exit visa application. The head of a laboratory at Chernogolovka is not likely to probe into much deeper causes, and it would seem strange to him to talk about the Russian political tradition as it affects scientific exchanges. He would hardly believe that the state's reluctance to allow its subjects to travel abroad goes back several centuries, yet that is precisely the case.

In the reign of Boris Godunov (1598–1605), seven Russian youths of good family were sent to foreign countries to study medicine. They were selected with the utmost care, not only for their ability but also for loyalty, which was more important, yet all seven of them failed to return. Many reasons may be suggested, but one is quite certain—they knew they would never be allowed abroad a second time. In Peter the Great's time fugitives from Russia were tracked down and brought home by force; under Catherine II they were publicly execrated. Thus for about four hundred years, and quite possibly longer, the Russian state has been waging a war against freedom-lovers. The state does not want its two-legged livestock to escape across the frontier, and in each new generation of slaves there are some who try to do exactly that. As the Marquis de Custine remarked in 1839: "Life in Russia is subject to such restrictions that everyone, it seems to me, secretly dreams of running away as far as his legs will carry him, but the dream is not destined to be fulfilled. The nobility are refused passports, and the peasants have no money."[8] About thirty years later, when the law against leaving the country had relaxed somewhat, Saltykov-Shchedrin wrote sadly: "One reform succeeds another, but Russian people still long to go abroad and are tormented with desire for foreign ways of life."[9]

It was not until late in the nineteenth century that Soviet scientists felt the need to travel for professional reasons, but when they did they at once encountered the traditional objection. "Where do you want to go? Why?" Even academicians traveling at public expense for research purposes had to obtain permission from the Tsar. There is a curious document, an "exit application" as it now should be called, relating to N. Gamel, a member of the Imperial Academy who in 1833 requested permission from Nicholas I to go to the United States to study "the system of telegraphic communication by means of a galvanic current." The Tsar consented, with the proviso that the academician must first sign a promise "not to consume human flesh by way of food, as is customary in America." This episode might be dismissed as a historical curiosity if it were not that, before and since Gamel's time, journeys by Russian scholars to foreign countries have always been subject to irksome and humiliating conditions. In the last thirty years of the Empire some relaxation set in, but as soon as Soviet power was established it restored the situation as it had been under Boris Godunov.

For the first three or four years the Bolsheviks allowed no one at all to leave the country. Maksim Gorky pleaded with Lenin on behalf of the academicians, but even he could not convince the dictator that scientific contacts were necessary. Journeys began to be allowed only after Pavlov addressed a sharp letter to Lenin demanding permission to go to England.[10]

"It cannot, alas, be concealed that not a few scientists succumbed to the temptation of using foreign travel as a means to escape from our country"—thus wrote, some years later, the aged S. F. Oldenburg, former permanent secretary of the Russian Academy of Sciences.[11] They were about the only people who had the opportunity of doing so: under Stalin, apart from a few engineers and production experts, no one was allowed abroad except diplomats, spies, and revolutionary agents. As for scientific contacts in the period of détente, our contemporaries can only envy Gamel for the fact that approval of his journey depended on the will of only one person. Nowadays it is such a complicated business that hardly anyone knows for certain just how many people have to give their okay before a Soviet professor or academician can leave the country.

The first time I discussed such trips at length was in Vladivostok, in the southeast, where I made the acquaintance of Galina L., a pleasant-faced, black-eyed young woman employed in a research institute. We got on to the subject of nationality, and she said with an embarrassed laugh: "All my friends are Jewish or half-Jewish, and my mother is very Jewish-looking; but God forbid that I should turn out to be one, or they would never let me go abroad, and I would lose my job as well." Galina is a biologist, and for the purpose of her work she needs to visit the countries bordering on the Pacific Ocean; but every visa application is itself like a voyage among dangerous reefs. On one occasion, as she was frankly told, permission was nearly withdrawn because her husband was corresponding with a foreign specialist.

Another restriction on foreign travel (like the rest, it is not mentioned in any law or regulation) is that a scientist cannot go twice to the same capitalist country: the only people who can do so are employees or agents of the KGB. This has led to the breakup of many fruitful relationships between, for example, Soviet biologists and those of the United States, Canada, and Sweden. "They write and invite us," said Galina, "and we can't explain our position to them. We can't write and say that we are all considered potential spies and deserters. You get a letter, and your cheeks burn with shame. That's why many of us have stopped writing to foreigners."

But let us suppose the trip is approved and everything goes through smoothly; what do they tell you at the official briefing? Always the same thing: the object of your journey is to give nothing away to foreigners, but to extract and wheedle everything you can from them. Many scientists are not prepared to accept such orders, but if they don't, they will not be sent abroad a second time.

There is one hazard that is particularly hard to avoid while one's application is under scrutiny. Everything may seem to be in apple-pie order: your party, professional, and social record is without reproach, your family situation is okay (i.e., family members will remain behind as hostages), you have no Jewish relatives and none living in foreign countries, nobody close to you has been subjected to repressive measures—and yet permission is suddenly refused! The blame, it turns out, lies with your well-meaning foreign colleagues, who have tried to facilitate your journey and speed up permission by telling the Soviet authorities that they will meet the whole cost of your stay in their country. The KGB reasons thus: "If they are so anxious to have this scientist of ours, they must intend to use him for some special purpose. We can't have that—he'd better stay at home." This has been known to happen even after permission for the journey was given, so that an unsuspecting scientist was hauled off a boat a few minutes before sailing time, or made to leave a plane already on the runway.

I had another conversation about such trips in the north, in Leningrad. Professor Izyaslav Petrovich Lapin, a psychopharmacologist at the Bekhterov Institute of Psychoneurological Research, is incomparably more famous than the humble Galina L. at Vladivostok. As soon as the student enters his laboratory, he realizes that the professor is a scientist of international rank. Not that the place itself is luxurious: it is a cramped, ill-arranged laboratory on the second floor of an annex to the main building, and the professor's office is far from roomy, but there is every indication that he is well known in the scientific world. The walls are covered with cordially inscribed photographs of such men as Julius Axelrod, the Nobel prize winner, Bernard Brodie, the founder of psychopharmacology, the world-famous José Delgado, and many others. Professor Lapin claims to have dozens of friends in foreign countries, and it is easy to believe this when one looks at his desk, with two large racks full of letters bearing foreign postmarks: one lot consists of requests for reprints of his articles, the other of invitations to attend symposia, give lectures, and so on.

Psychopharmacology is a young, rapidly developing science on the borders of psychiatry, physiology, and chemistry. Discoveries crop up like mushrooms; ideas crystallize with tremendous speed, and are of interest to millions of people. Not surprisingly, specialists in this science are keen to meet one another. Professor Lapin's articles in learned journals contain new and exciting information, and one would think that nothing should stand in the way of his contacting foreign colleagues. However, the authorities take a different view. Lapin is one of seven members of the Pharmacological Section of the Committee of the International Psychiatric Association, which has been meeting twice a year for seven years. Eminent scientists come from Germany, France, Britain, and the United States, as well as Mexico and Canada; only the Russian seat is vacant on each occasion. Perhaps Professor Lapin does not like large public gatherings? Then let him come to Milan, to the three-week seminar of Italian psychopharmacologists, to which Dr. Silvio Garottini, the chief Italian expert in this field, has invited him for the third year in succession. Or why does he not go to Boston, where he has been invited several times by Dr. J. Mendelson, director of the Center for the Prevention of Alcoholism and Drug Addiction at McLean Hospital? But no, it is no use—Professor Lapin will not go. Every day he receives six or seven letters from abroad, yet he stays at home. Is he too busy, or is his health the problem?

"Every time I get an invitation," the professor told me, "I feel like a cripple being invited by his athlete friends to race around the track. I feel helpless and deeply humiliated, and the sense of enforced inferiority literally makes me ill."

That is how the professor feels. But what actually happens when he receives an invitation? He asks the director of the Institute for an interview on a personal matter, and if he is granted an audience he shows his superior the letter. The director cannot read foreign languages, so the laboratory head, who knows English, German, Italian, Polish, and Hungarian, translates it to him. "But why do you want to do this?" the director generally replies. "Why Milan and Boston all of a sudden? Wouldn't it be better to start with Bratislava?" The specialist would be happy to go to Ryazan or anywhere else, but the invitation happens to be from Milan. If the director is in a good mood he will make a vague gesture signifying: "All right, we'll give it a try."

This is the signal for the first round of activity. A photocopy of the invitation, with a Russian translation and a covering letter from the Institute, is sent to the Ministry of Health in Moscow, where the Foreign Department will consider whether it is desirable for Professor Lapin to go to Milan. Leningrad settles down to wait—a month, two months, three months. The professor applies once more to the director, who says: "Better wait a bit longer, we don't want to annoy them." After six months have passed the director goes to Moscow and cautiously inquires how matters stand. He is told that the position is not entirely clear. By the time it does become clear it is too late anyway—the symposium has already taken place.

It may happen, however, that a response will emerge from the depths of the Ministry: "Please send an exit application." This is joyful news—it guarantees nothing, of course, but at least some official is prepared to consider a formal request. At this second stage it is necessary to procure and certify twelve documents: a health certificate signed by six doctors, a record of conduct with three signatures and the stamp of the party's regional committee, and so on, and so on. The applicant must also have a personal interview with the regional committee and also the party committee at the Institute.

These interviews are the most crucial part of the exercise. Five or six elderly officials, generally veterans of Stalin's day, ply the professor with questions. "Why did the United States embark on a political rapprochement with the Soviet Union?" The correct answer is: "Under the pressure of the Soviet Government's peace-loving policy." If you don't know this formula, you are told that you are not mature enough to be sent abroad. The professor of pharmacology also has to answer questions like: "What does dialectical materialism understand by 'chance'?" "Who is Alvaro Cunhal?" "What is the peculiarity of the present stage of peaceful coexistence?"

Lapin, in fact, was unable to satisfy the examiners on these questions, and so he was not allowed to go. He was at least lucky not to be mocked, which can also happen.

A student of geography in Leningrad, who was supposed to go on a foreign trip as part of his university course, was asked: "For you to spend three days in Dover you have to go through a probation of seven stages lasting three months, whereas a Frenchman has only to buy a ticket on the cross-Channel ferry. What would you reply if they put that to you?"

The student answered that he would not talk to foreigners about such matters, but apparently this was the wrong reaction, as he was not allowed to go. . . .

The party officials may question applicants on every subject under the sun, for instance: "What are the scientific qualifications of the Emperor of Japan?" "How many tons of steel were smelted in the USSR in 1970?" and even: "Why did you divorce your wife?" Those with experience say that what matters is not so much the content of the questions and answers as the applicant's behavior at the interview. If he replies boldly and cheerfully, he gets a good mark; if he is hesitant or gloomy or appears vexed by the interrogation, it counts heavily against him. The most dangerous thing is to show the slightest trace of irony or sarcasm—this will be neither forgiven nor forgotten. But the main point of the whole game is to humiliate the applicant who may, in a few days' time, actually be at large in a free country. True, the final decision on the application is not taken by the party committees but by the KGB, to whom the papers are forwarded. But the party officials have the function of reminding the professor that he is not his own master and must not step out of line.

As long as the application is being processed, the aspiring traveler cannot reply either yes or no to his foreign colleagues. Officially he is advised not to write to them at all until a decision is made; then, if it is negative, he must say that he is sick or too busy to travel. Compliance with this rule is one more test of obedience.

Professor Lapin was in despair. He was not allowed to go to Geneva or the United States, or to the sixth World Congress on psychopharmacology, held at Honolulu in September 1977. All that is not for him. In some way that he does not understand, he has annoyed the authorities and the road to capitalist countries is closed to him. But surely he might go to Poland, a fraternal socialist state, where a symposium on his subject was being held at Krakow? No, that was forbidden (1973); and in 1974 he was refused leave to visit a scientist friend in Warsaw. In 1975 he went to Moscow to plead his case at 6 Ogarev Street, the Ministry of Internal Affairs, where a KGB general made no secret of his surprise at Lapin's complaint. "If I had my way," the general added, "no professors or doctors would be allowed to travel at all. Each of them has a hundred friends and pupils all over the place—are we to let them visit a hundred countries?"

Gustave Doré, who illustrated Custine's book over a century ago, depicted a Russian citizen crossing the border on the outgoing journey with the friskiness of a young colt, and returning home with heavy tread as though he were being led to the slaughter. The same contrast was noticed by many writers and travelers, including the revolutionary Ogarev for whom the Ministry's address is named. But whatever treatment a returning Russian citizen might have expected in the nineteenth century, he could not have imagined what recently happened at a Moscow research institute. Two research officers, candidates of science, were preparing to attend an international conference on ferromagnetic materials, a specialized and complicated subject on which they were the only available experts. But at the last moment they were forbidden to go and were replaced by two ministry officials. The officials knew little about the subject, but one of them was able to read a paper written by the experts and even stammer a few answers to questions. Then, having had their fill of foreign delights, the officials returned to Moscow laden with conference papers. The material was far above their heads, but to their alarm the Academic Council of the Institute called for a report on what they had seen and heard at the conference. After a moment's perplexity they called in the two specialists who had been forbidden to go, and ordered them to write the report on the basis of the documents. Was anyone ashamed or embarrassed? Not at all. The officials delivered the report with a straight face, and the Academic Council did not turn a hair. Just an ordinary instance of "manipulated science."

But perhaps the people I have written about—Dr. Likhtenshtein of Chernogolovka, Professor Lapin of Leningrad, Professor Isayev of Samarkand, and the Moscow ferromagnetic specialists—were all atypical failures, and the great body of scientists who have professional reasons for traveling abroad are in fact allowed to go, so that scientific contacts are kept up and the mental balance of researchers does not suffer? This, at any rate, is the official version. . . . I could not obtain information about doctors going abroad, but I found out a few facts about the Academy of Sciences of the USSR. Someone who served on its Presidium for many years told me that out of every hundred scientists who apply to be sent abroad, not more than ten receive permission; the rest are refused, or their papers are not dealt with in time, or the file is "lost," etc.—whatever the excuse, 90 percent do not succeed in going.

As I remarked to this informant, it must require a large staff of officials to defeat so many hopes, to stop so many mouths and ears. He agreed, and drew my attention to the Academy's annual register for 1972, from which it appeared that some 250 people in the Academy building had the official duty of regulating scientific contacts; the foreign relations department numbers 150. I was given a detailed account of how this army of officials works, how it tricks, deceives, and humiliates those who are to be prevented from traveling. The picture was a fairly monotonous one, but two details are of interest. It appears that where a member of the Academy is concerned there are eight stages of official clearance to be gone through, as the application also goes to the science and propaganda departments of the Central Committee of the Communist Party of the Soviet Union, in addition to a long inspection by the KGB. Having enumerated all the hazards and pitfalls, the Presidium official suddenly said in a serious tone: "We are doing our best to put an end to that nonsense." For a moment I thought he meant that the academicians were actually defending their right to travel and meet foreign colleagues whenever they chose. But it turned out that the Presidium was only fighting for the principle that travelers should receive clearance a day or two before their departure instead of at the very last moment. The academicians are fighting for this, but so far without success.

I questioned my informant about the effect of détente on scientific contacts. Apparently the question of a relaxation was seriously discussed as far back as the summer of 1973, shortly before Brezhnev's trip to the United States, when the Soviet Academy sent a group of physicists to America to prepare the way for a high-level agreement. When the emissaries returned there was a discussion in the Presidium, attended by KGB General Stepan Gavrilovich Korneyev, head of the Academy's external relations department. The Americans had put forward a radical proposal whereby ten times as many Soviet physicists as before would work in American laboratories and vice versa, and for twelve months instead of two. General Korneyev objected, however, that if the Americans came for a year they would want to bring their families; Soviet families would have to be allowed to go to America, and what was to stop them from running away? No, the Americans could bring their families to Russia if they liked, but we would send ours without families, and for four months only.

This seemed a wise solution, but then someone objected that it would mean not ten, but thirty Soviet scientists going to America annually. At this point an academician who had not spoken until then said with a sigh: "Goodness, what a lot of work this increased exchange is going to give the FBI, not to mention our own security people!"

Korneyev was undaunted, however. "Don't you worry," he replied, "if the contacts increase we'll be given extra staff to deal with them."

After this reassurance the discussion went on more cheerfully, and the academicians soon approved a rational plan.

Korneyev is an extremely colorful figure in the world of Soviet science. His successes in discouraging contacts of all kinds deserve to be the object of a special study. The younger academicians regard him as the embodiment of repression, but he himself takes a different view. Having spent several decades among academicians he developed scholarly ambitions himself, and presented a thesis in the early seventies for the degree of candidate of historical sciences. This work, greeted with acclaim by the Academic Council of the Oriental Institute of the Academy of Sciences, was titled, believe it or not, "The International Links of the Academy of Sciences of the USSR." I went to the Oriental Institute to find out what new ideas the general had concerning scientific exchanges, but a professor whom I knew advised me not to probe too deeply. "The members of the Academic Council also want to travel abroad," he reminded me. "Anyway this is old stuff, the general is now working up to a doctor's degree."

The last paragraphs of this chapter were written at the end of May 1976. One of my literary subjects, the pharmacologist Izrail Itskovich Brekhman,[12] had come to Moscow two weeks before. At the end of the previous year he had been invited to speak at an international symposium held in Singapore and comprising representatives from Western Pacific countries. His own specialty was the pharmacology of the araliaceae (ginseng, eleutherococci). From Vladivostok to Singapore is not so far, as the crow flies, but a Soviet scientist, no matter where he is stationed, must begin every foreign trip from Moscow. The professor, having obtained all necessary clearances, had therefore come six thousand miles to collect his documents. A punctilious man, he arrived at the external relations department at exactly 3 P.M. as instructed. Everything was in order: passport, visa, money, plane tickets. Yet Professor Brekhman never got to Singapore. Two hours earlier, a high official of the Academy had given orders that he was not to go. No explanations, of course, and no use protesting or complaining, as the flight via Delhi was due to start in a few hours' time. So the professor traveled back another six thousand miles to Vladivostok.

At Singapore, no doubt, a scene took place which has occurred at many another conference. At the appointed time, the chairman of the conference announced that Professor Brekhman of the USSR would read his paper. After waiting a moment or two and seeing that the Russian pharmacologist was not present, the chairman suspended the session for half an hour with the words "Russian time, gentlemen—Russian time." I don't know who coined this phrase in English, but it is often heard at international scientific gatherings when, as so often happens, the Russians unaccountably do not turn up.

Are we to suppose that Russian scientists are kept from traveling abroad merely for fear that they might not return? No, the reason is deeper than that. The Soviet leaders do not want any contacts at all between Soviet citizens and foreigners. Constant interference at all levels is designed to prevent private contacts and unauthorized friendship. In Stalin's time those who kept in touch with foreigners or corresponded with foreign countries were arrested and sent to labor camps. This no longer happens, but all officials, including those who deal with science, have an ingrained hatred of contacts with the outside world.

In May 1976 a symposium on the immunology of tumors was held in Moscow under the Soviet-American agreement on medical cooperation. Afterward the leader of the American delegation, Dr. William Terry, director of the National Cancer Center, complained formally to the Soviet Minister of Health that Academician O. Baroyan, director of the Gamaleya Institute of Microbiology and Immunology in Moscow, had done his best to hinder contacts between Soviet and American immunologists. I asked a group of doctors of science at the Institute whether Terry's complaint was justified. Yes, they said, Baroyan (who was also a KGB colonel) had prevented any discussion going on in his laboratory. He made no secret of his dislike of Soviet-American scientific contacts, of anything which improved mutual knowledge and appreciation of one another's work. "There's another reason, too. A Soviet scientist who becomes known abroad is more independent of his superiors, of Baroyan himself, than one who is not known. Contacts with foreigners make us freer, and that is anathema to officials like him."

One might conclude the account of international scientific contacts at this point, but there is a further aspect of the problem. The delegates at symposia who sit patiently smoking cigarettes and making jokes about "Russian time" as they wait for the next paper to be read might take a more serious view of this antiquated situation. After all, Korneyev's officials are not only insulting Russian physicists, chemists, and biologists; they are spitting in the

face of the whole scientific world. They are doing an injury to Soviet and world science, but an infinitely greater injury to human
dignity and to the moral sense of the five-million-strong international republic of thinkers, discoverers, and inventors. I have only
once heard of a letter from a British or American physicist being
published in *Nature,* expressing indignation at the fact that Soviet
delegates to international meetings are unknown, colorless individuals and not scientists whose work is known to their Western colleagues. This letter was rightly acclaimed by Soviet scientists, but
why have there been so few like it? Why is the world scientific
community so shy at speaking up for the freedom of scientific contacts?

Not long ago a group of Soviet chemists who turned up for a
symposium in the United States proved, once again, to be made up
of different individuals from those who were expected. The organizers wished to protest and turn the matter into an international
issue, but the leaders of the Soviet group dissuaded them with the
argument that it would only make things worse. As it was, a few
genuine scientists were allowed to attend symposia, but if people
started to protest there would be none. This specious argument
convinced the Americans, who remained quiet. Freedom-lovers as
they were, they feared their actions might make the Russians even
less free.

So no protests at any cost! True, the world Psychiatric Association, meeting in Geneva and finding Professor Lapin absent as
usual, sent a sympathetic letter to Leningrad: "We are sitting in
the hall in which you were to have addressed us this morning, and
we send you this message of solidarity." Solidarity is fine, but do
the members of the association know that now, after the Helsinki
conference, it is harder than ever for scientists to leave the USSR?
Do they know that a new rule requires Professor Lapin to obtain
not only the consent of the party committee and the director but
also the unanimous approval of all party members in the Institute,
who have to express their confidence in him at a special meeting?
In other words, if a porter or elevator girl at the Institute,
prompted by the party committee, raises any objection to the professor's journey (and why shouldn't they?—no one sends *them* on
foreign trips), he can say good-bye to any hope of meeting his
American and European colleagues.

That, dear foreign colleagues of Professor Lapin, is the reality behind jokes about "Russian time." Does it not seem to you insulting that science, which is one and indivisible, should be split into two unequal parts? Are you not distressed by the fate of your comrades whose words, as Mandelshtam put it, are "inaudible ten feet away"? Would it not be worthwhile for you, now and again, to speak out loud and clear? Voltaire once said that the time of a learned man is the most valuable time in the world. Must we not recognize that "Russian time" is not only time lost by Russian scientists but a loss to all humanity?

CHAPTER 6

A TOWER OF BABEL WITH A FIVE-POINTED STAR

"What could be more illogical than Tashkent resolving firmly to remain illiterate, when it is already covered by a veneer of civilization?"*

About twenty years ago, in Tashkent, I made the acquaintance of Mikhail Sionovich Sofiev, a doctor of medical sciences. The doctor's race and local origin were indicated by his olive skin and Jewish patronymic: he was a Jew from Bukhara and had spent his whole life in Uzbekistan, devoting himself to the study of epidemic diseases in Central Asia. His account of the fight against Turkestan ulcers, dracunculosis, malaria, and other scourges was more exciting than any detective story. His conversation was the more fascinating because of his wealth of detailed knowledge, based on personal experience and practical work. In addition he was a good and kindly man, which is an adornment to scientists as to other people.

One day we were talking peacefully in his office in the epidemiological department of the Tashkent Medical Institute when there was a knock at the door, or rather a scraping noise as though a dog were asking to be let in. To my surprise Sofiev took no notice, even when the noise was repeated; only an irritable twitch of the eyebrow showed that he realized that someone was demanding attention. Then the door opened gradually; a small, swarthy hand came into view, and behind it the face of a young woman. Finally three girls, wearing long oriental dresses and flowered kerchiefs, squeezed through the aperture and stood motionless in the doorway behind Sofiev, who pointedly ignored

* M. E. Saltykov-Shchedrin, *Complete Works,* Vol. 10, p. 30.

them. One girl began to sniff and whimper, and the others followed suit. Without turning his head, Sofiev asked them something in Uzbek. The girls replied through their tears, and sobbed even louder.

"I expect you're sorry for them?" said Sofiev to me. I replied that of course, when young creatures seem so distressed . . .

"Those young creatures are simply idlers and scroungers," Sofiev exclaimed. "All they want from me is a mark in their record book, even though they haven't taken the exam or done a lick of work."

From his further explanation it appeared that these students were actually demanding that the professor give them a "satisfactory" or even a "good" mark in a subject they knew nothing whatsoever about, namely epidemic infectious diseases. The other teachers, they argued logically, had already given them marks for subjects in which they had not passed any test, and now they needed one for epidemiology. Without it, according to regulations, they could not return their bed linen to the students' hostel as a preliminary to going on vacation. They were homesick and wanted to be off, so would the professor please give them the necessary mark? What difference did it make to him? It was urgent, too, as the bursar himself was going away soon and the linen had to be returned that day.

The professor held his head in his hands and spoke to them angrily.

"Can't you understand," he said, "that you'll be sent as doctors to the villages, and you know nothing about the diseases you'll have to deal with. What's to become of your patients? . . . Go off and do some reading!" The girls went on sobbing and muttering about pillows and blankets. "Study, damn it!" roared Sofiev, and the girls tumbled out into the corridor. I could not help laughing, but Mikhail Sionovich was not amused. Could he really have been upset by this trifling occurrence?

"The girls say the other teachers have given them marks, so presumably they've studied *something?*" I asked.

"They haven't studied a thing, and they're quite incapable of doing so. Their school education is so backward that they have no idea what the Institute lectures are about—they simply don't understand them. They haven't taken the exam in anatomy, or physiology, or microbiology."

"And yet they are allowed to move on to the next year's course?"

"Yes, because they are natives of the Karakalpak Autonomous Region."

At last I realized why Dr. Sofiev was so angry. Like other members of the Institute faculty, he was obliged by regulations to give passing marks to students from Karakalpakia, an autonomous region of Uzbekistan. Indeed, many lecturers also gave unjustified grades to Uzbek students of all kinds, and had been doing so for decades. Such is the official policy of the USSR with respect to the borderland peoples. Young men and women from primitive villages, with scarcely any education, are reluctant to study in the cities: they have to be begged, cajoled, and practically forced to do so. They are given scholarships and free lodging, and are assured of passing marks whether they study well or not. The philosophy behind this strange proceeding is that all nations in the brotherly family of the USSR are equal, and that all of them therefore can and must have their own intelligentsia, their own doctors, engineers, writers, and scholars.

"Those three girls will be doctors soon," said Sofiev, shaking his head sadly, "but God preserve you or anyone else from having to go to them for treatment."

Meanwhile the girls in kerchiefs reappeared in the doorway, no longer crying but smiling triumphantly and brandishing their grade books. All the whining and sobbing and scratching at the door had been mere playacting, almost a matter of routine. Other instructors did not keep the girls waiting so long, but awarded grades for nonexistent attainments, on demand. Only Sofiev could not reconcile himself to doing this, but the girls had brought him into line, knowing as they did that a passing mark was literally theirs by right of birth.

As I took leave of Sofiev he told me that we could not meet for a few days, as he was going on a trip. "Where to?" I asked.

"To Karakalpakia," he replied with a wry smile, "to enroll the next quota of medical students, preferably girls—the percentage has to be kept up."

Professor Sofiev died many years ago, but the performance I witnessed in his office still goes on all over the Central Asian republics, in Azerbaijan, and in the national and autonomous republics of the Caucasus, Siberia, and the middle Volga. Higher education of the "Tashkent" variety is still the birthright of thousands of young Turkmenians, Azerbaijanis, Uzbeks, Karakalpaks, Maris, Udmurts, Kirgizes, and Kazakhs, nor should the youth of Moldavia and Belorussia be excluded from the list. A hundred years ago Saltykov-Shchedrin took Tashkent as a symbol of primitive backwardness in the Russian Empire, but this is not a purely local phenomenon: it stems from a principle observed in the Soviet Union for the past fifty years, that of creating "national intelligentsias."

The mass production of these intelligentsias in the borderlands is above all a matter of politics. No other social measure could have had such a dazzling propaganda effect as the sudden burgeoning of schools, institutes, doctors, scientists, and scholars in Siberia, Central Asia, and the Caucasus. As we have seen earlier,[1] the authorities did not care in the least whether the newly fledged doctors, engineers, and teachers were of good or bad quality: they merely wanted the maximum number at the lowest cost. The great object was that when an international congress was held at Tashkent in the presence of delegates from India, Cambodia, and Dahomey, the Presidium should include a dozen or so professors and academicians of Uzbek, Tajik, and Turkmen nationality. The Asian and African visitors, enchanted by this sight, would regard the fatherland of socialism as an earthly paradise. They could not have any idea what it really meant to be an academician or a doctor of sciences from Turkmenia or Kazakhstan, any more than they could imagine the true quality of the cadres of Soviet teachers, physicians, and agronomists.

I am far from suggesting that the native peoples of Asia, the Caucasus, Belorussia, and Moldavia are devoid of intellectual gifts; but I do say that hundreds of thousands of young people from these areas are systematically deceived. What is presented to them as higher education would not deserve that name even in Moscow or Leningrad. The deception, however, does not distress them overmuch. As the word goes out from Moscow to the minority republics to produce "national cadres," and as secret instructions are given to see that their students get good marks at all costs, the students soon catch on to the fact that it is easy to complete a "simplified" university course and that a degree confers enormous advantages. Gifted and ungifted students pour into scientific institutes, knowing that their ability or lack of it makes little or no difference as far as graduation is concerned.

In the early forties, having ensured an adequate production of graduates in the southeastern republics, Moscow embarked on a second round of "intellectualization" in these areas. The years between 1941 and 1951 saw the establishment of ten republican academies supervising over 150 research institutes. No one regarded these academies as centers of creative learning; the emphasis was strictly political. The new centers were to exemplify another virtue of socialism, inasmuch as their scientific activity would be not only controlled but supercontrolled. In this way academies sprang up in areas which, a few years before, had counted barely half a dozen holders of higher degrees. These select few had been dubbed academicians on the spot, and their half-educated pupils were rapidly enrolled as candidates and doctors of sciences. The loyalty and amenability of the new recruits was beyond question: the ex-villagers with diplomas in their pockets would do anything that was asked of them.

The new cohorts of learning were commanded in the first place by scholars who had migrated from Moscow and Leningrad on account of the war and were anxious to make a niche for themselves, socially and materially, in the depths of Central Asia. Their personal interests thus coincided with political and ideological ones. E. N. Pavlovsky, a parasitologist, became the father of the Tajik Academy of Sciences, while the helminthologist K. I. Scryabin founded the Academy of Sciences at Frunze, capital of the Kirgiz SSR. Other migrant scholars from Russia proper found cozy berths in the warm climate of Azerbaijan, Kazakhstan, Uzbekistan, and Georgia. During the war years, academies sprang up like mushrooms. After 1945 their founders returned whence they had come, but the institutes continued to multiply.

In the year when Evgeny Nikanorovich Pavlovsky left besieged Leningrad and betook himself to the Tajik capital of Dushanbe (then called Stalinabad), there were scarcely three or four hundred people in the republic who had any pretension to be called scholars, and all of them came from outside. Higher degrees were only known to the Tajiks by hearsay, and academicians not at all. But everything changed in a twinkling when Pavlovsky first established a local branch of the USSR Academy of Sciences, and then a fully fledged republican Academy. Well-paid jobs were suddenly plentiful: as the proverb says, a holy place is not empty for long. Between 1950 and 1956 the number of scientists in Tajikistan multiplied fivefold, and in the next decade it doubled again. At the present time there are two thousand candidates of sciences, about two hundred doctors, twenty-two academicians, and nineteen corresponding members of the republican Academy; the vast majority of all these are Tajiks. In the other Asian republics the position is similar. Just as the birthrate in Central Asia is more than twice as high as in the RSFSR or the Ukraine, so the mass production of scientists has proceeded at a rate which leaves the Western parts of the USSR far behind.

However, in order to be assured of a quiet and comfortable life, a scientist, wherever he lives, must in due course present a doctoral thesis. What is the standard of these compositions, the open sesame to the happy land of science? In academic circles in Moscow and Leningrad one may hear many anecdotes about the illiterate scripts received from Central Asian graduates; but let us hear what some experts of the Higher Examinations Board have to say.

Dr. F.B., geneticist:

The chief spawning grounds of candidates and doctors are the Academies of Turkmenia, Azerbaijan, and Tajikistan. The number of theses received from there is absolutely frightening. The applicants not only don't understand the scientific material they are handling, they can't even write grammatical Russian. The law is that scientific and technical theses have to be in Russian, but the stuff they write at Baku or Ashkhabad would not have been recognized as Russian by Tolstoy or Chekhov. But if we presume to fail these candidates they complain to the republican party organization, which passes on the complaint to the Central Committee in Moscow, and the Board receives instructions to accept the theses no matter how illiterate they are.

There are many cases of fraud, too: a thesis will be written by one person and presented by another. For this reason the Board has more than once forbidden certain Academic Councils in the republics to accept theses for presentation. A one-year ban of this kind was imposed on the Joint Academic Council of Biological Institutes of the Azerbaijan Academy, and on some Academic Councils in Tajikistan. But, to judge by results, the output of these "dissertation factories" has not been affected in the slightest.

Dr. V.I., cytologist:

In 1974–75, when the Board suspended work for a year and a half owing to reorganization, files accumulated in our offices in stacks five or six feet high. Most of the theses I had to plow through were from Azerbaijan, and they were extremely poor. However lenient I try to be, I have to fail 15 or 20 percent of the work submitted. The appalling state of corruption in the republics is evident from a glance at the "thesis file" [containing the text of the applicant's oral statement and answers to questions, reports by his supervisor and examiner, recommendation of the Academic Council, etc.]. As often as not the "examiner's report" is really written by the candidate himself. The farther southeast you get, the more flowery and superficial are the recommendations. Everyone praises the candidate to the skies. Many of the reports are obviously written on a "friendly" basis by specialists in some quite different subject.

How does the candidate repay his examiners for their kindness? If the thesis is accepted, the examiner who has seen it through will be invited to Baku or Tbilisi, wined and dined and given a free holiday on the Black Sea or the Caspian at the candidate's expense. Sometimes repayment is in kind. An examiner from Moscow will pass a thesis concocted in Baku or Tbilisi, and later on a pupil of his will present a thesis in one of these republics, knowing well that he is among friends and that they won't let him down—nor do they.

No one in Russia nowadays is astonished or indignant at the widespread fraud practiced in academic matters, bolstered as it is by politics and ideology. What would once have been considered shameful is now a commonplace in the minority republics, where candidates and doctors see no reason to hide the way in which they obtained their degrees. At a conference at the Alma Ata Medical Institute, I attended an interesting lecture on the effect of red light on the regenerative properties of nervous tissue. The speaker was a short man of distinctly Korean appearance, but according to the program the authors of the paper were Rakhishev, doctor of medical sciences, and Tsoy, a physician without a higher degree. I sent a message saying that as the correspondent of a Moscow newspaper I would like to meet the authors and discuss their investigations with them. In response I was introduced to Professor Alshimbay Rakhishevich Rakhishev—an enormous, broad-shouldered man in a stylish gray suit, with a florid countenance and the plump figure of one who liked his food and drink.

Professor Rakhishev entertained me for half an hour with the story of his life, intermingled with the political history of Kazakhstan, the elements of biology, and his own academic philosophy.

He was a shepherd's son, born after the Revolution, and had been working on a collective farm when some kindly Russian folk persuaded him to enroll in the Karaganda Medical Institute. He obtained wretched grades in his first year and begged to be allowed back to the steppes. But the director had orders to produce physicians of the local nationality, and would not let him go. From then on he was carefully given good grades and, although he knew nothing of anatomy or physiology, was accepted for postgraduate studies, after which he never looked back. He was appointed an instructor in anatomy at the newly opened Medical Institute at Tselinograd (formerly Akmolinsk), where in two years he became a candidate of sciences. He then went to Leningrad and returned two years later with a doctor's degree. His successful career was noticed; he became assistant head of the anatomy department and prorector of the Alma Ata Medical Institute. He spoke about his career with the air of a man who has always been lucky and expects to go on being so. As for scientific problems, he discussed these on the level of a medical orderly's manual and clearly was not at all embarrassed by the paucity of his knowledge.

I asked Rakhishev about his coauthor, Tsoy. "Oh," he replied carelessly, "he's a Korean working in my department."

"No doubt the idea of the experiment is yours, and Tsoy is just the executant?" I inquired with secret amusement, expecting Rakhishev to show even greater complacency. But his black eyes gleamed curiously as he answered: "The experiment was Tsoy's idea, but the department of normal anatomy belongs to me."

No, Professor Rakhishev was no fool. He did not know any anatomy, but he knew perfectly well that knowledge was one thing and the ownership of it was another. . . .

The system whereby one person does the work and another takes the credit is so prevalent in the Asian republics that the rank and file make no bones about admitting it. A Moscow mathematician told me laughingly of a mishap he had suffered. Asked to write a thesis for a "customer" in Georgia, he suggested a fee of 5,000 rubles. He heard nothing more of the idea, but some months later it turned out that the Georgian had ordered his thesis from another specialist in the same institute, who had gained his confidence by asking 15,000 rubles—the customer evidently thought that the higher the price, the better would be the quality of the goods.

Here are some lines from a letter I received from Uzbekistan in the fall of 1972:

> . . . I thought that when you left our town you seemed rather vexed, even angry. Why? Can it be because I told you that I had written seven theses for local Uzbeks? What choice did I have? The people I wrote them for are highly placed, and if I hadn't done it, someone else would have. As it is, they let me get on with my own work and help me carry through measures to prevent epidemics. It's the only way you can get anything done in Uzbekistan. If it wasn't for those theses, I and my scientific ideas could go to hell. So don't be angry—*c'est la vie,* as they say nowadays!

The writer of that letter is a helminthologist who has done important theoretical and practical work on the elimination of an intestinal parasite which infested children's homes. I had been arguing with him during an excellent dinner at his house. A guest, of course, has no business to start chiding his host for debasing the coinage of science and filling the profession with incompetents. My friend returned a soft answer at the time, but wrote to me afterward explaining the simple reason why he wrote theses to order. If he did not, it would mean a setback to his work against intestinal infection. What answer is there to such an honest explanation?

The system of "science for hire" sometimes leads to dramatic situations. In Samarkand I knew a zoologist, an expert on rodents, who had lived many years in Central Asia and knew well the local rule that before writing your own doctoral thesis you should write one for your official superior. He did so, and expected that he could then make his own way forward, but no—he had forgotten the director of the Institute. This man, an Uzbek, was only a candidate of sciences and did not want his subordinate, a Russian, to become a doctor. This was a typically oriental reaction: a higher degree, in the East, is not so much a sign of knowledge as a symbol of power. A doctor has more clout than a candidate, and a candidate more than a junior research officer. To prevent his Russian subordinate from getting above himself, the director resorted to an equally oriental expedient: he told two members of his staff to break into the zoologist's laboratory and steal the collection of rodents on which his doctoral thesis was based. However, the zoologist's friends got wind of the plan, armed themselves with double-barreled sporting rifles, and kept a round-the-clock watch in the laboratory, at the same time letting it be known that they would shoot anyone who approached it with evil intent. They mounted guard in this way until their friend had successfully presented his thesis. No blood was shed, but, knowing local ways, it is not hard to imagine a more violent outcome to the story.

Having visited Azerbaijan, Uzbekistan, Kazakhstan, and Kirgiz at intervals during the past two decades, I observed the gradual progress of the idea, which is now universal in those parts, that a local scientist should not have to write his own doctoral thesis. It is not his job to wrestle with scientific argument. He can easily get someone to prepare the necessary 250–300 typed pages: they are his due, and if it comes to that point he can simply buy a thesis, complete with the regulation synopsis. Anyone who does the work for himself nowadays is a laughingstock to the natives of Tashkent, Dushanbe, and Baku, just as nineteenth-century Moscow ladies would have laughed at one of their number paying visits without a carriage and without servants.

Another curious fact is that in the whole course of my travels in the southeastern republics, none of the Azerbaijanis, Uzbeks, or Kazakhs to whom I talked ever said a word about the idea of scientific responsibility—it simply did not enter their minds. Odd as it may seem, it does not worry them in the least that ignoramuses should be heads of laboratories, hospitals, or engineering offices. Their view is that a candidate's degree should be an absolute protection against any future professional anxiety.

A distant relative of mine, a surgeon and an expert in the cure of strabismus in children, works in the ophthalmic department of the Medical Institute at Kharkov. She told me once of a young woman doctor from Kirgiz who had come to the department as a probationer but had made no attempt to do any work. Soon she would be returning home and expected to operate on child patients, yet she scarcely knew how to hold a scalpel. The surgeon spoke so indignantly and with such evident concern that I decided to have a word with the doctor from Kirgiz. I went into the interns' room, and among a dozen or more men and women in white smocks I immediately noticed a trim little woman with a doll-like face, wearing a smock that fitted as perfectly as if she had been born in it. We got to talking, and after some general questions I asked how she liked her chief and director of studies. Yes, she replied, the surgeon was a good sort of woman, but she pestered the staff with too many instructions.

"She is worried about you," I rejoined, "because you don't seem much interested in operational practice. After all, when you get home you'll have to operate on children."

"Oh, she needn't be anxious about that," replied the Kirgiz woman in excellent Russian. "I shan't ever have to treat anyone myself—never! All I need is a candidate's degree—once I've got that, I'll be chief medical officer as a matter of course."

This was said with such a charming smile that I could think of no reply. It is not every day that a surgeon who is about to take a higher degree tells you that he or she has no intention of ever practicing. The confession in its naïve sincerity had a captivating effect, one of innocence.

It should not be thought that no one in the Central Asian republics is perturbed by the current state of affairs, especially in the medical sphere. The lack of responsibility among native doctors and their low professional competence are noticed not only by European Russians but by the most honest and sensible of the local administrators. In Bukhara I made friends with Shovkat Gulyamovich Yuldashev, a forty-year-old neuropathologist and a man of real culture, with a lively interest in the history of his country. He is head of the Regional Health Department, covering a huge territory which includes a sizable part of the Kyzyl Kum. He has shown great ability and energy in this job, and suppressed several outbreaks of plague and cholera in Bukhara and the countryside. During one of these, I was told, the senior local official, the secretary of the Regional Committee of the CPSU, said to Yuldashev: "You're in charge of the city now, tell us what to do and we'll obey you." I kept a note of my conversations with Yuldashev, who said to me in the fall of 1972: "We're not happy with the situation as regards local doctors—the quality gets worse every year. There are more and more people with degrees, but no one to look after patients. There are six doctors under me in the neurological department of the local hospital; two of them have candi-

12.   Academician Lev Artsimovich (1909–73). His statement that "science lies in the hands of the Soviet state and is sheltered by their protecting warmth" became the epigraph for this book.

13.   Mstislav Keldysh (1911–78) became a member of the Academy of Sciences of the USSR in 1943, and president of the Academy in 1961. A leading theoretician in the Soviet rocket-building industry, he actively assisted in the militarization of the Academy.

14.   During the time Aleksandr Sidorenko (born 1917) was Minister of Geology, he managed to become an academician.

15. Academician Aleksandr Frumkin (1895–1978). In spite of his great scientific achievements and position as a director of the Institute of Electrochemistry, he was never allowed to forget that he was a Jew.

16. Academician-historian Sergei Platonov was one of the first academicians to be condemned by the Soviet officials for his views. He died in exile in 1933.

17.   Although academician Aleksandr Aleksandrov (born 1912) is a mathematician, his favorite topic at public appearances is justifying the Soviet regime. He signed the Communist Party denunciation of the dissident scientists of Academgorodok, in Novosibirsk. He has been a member of the Communist Party since 1951 and has been decorated with many orders.

18.   Academician-chemist Nikolai Zhavoronkov (born 1907) has been a member of the Communist Party since 1939. A "Hero of Socialist Labor," he supported the construction of a military plant near Lake Baikal, and sanctioned the use of this unique lake for industrial waste disposal.

19.   A Nobel Prize winner, academician-physicist Nikolai Basov (born 1922) is famous among the members of the Academy of Sciences for his cynical amorality. His colleagues used to call him a "smiling shark."

20. The geneticist and academician Nikolai Dubinin (born 1907) was persecuted during the Stalinist and Khrushchev eras. In 1973, by order of the Communist Party, he published a book called *Eternal Movement*, in which he defended Lysenko's dictatorship and Stalin's persecution of Soviet biologists. He was awarded the title "Hero of Socialist Labor."

21. and 22. Academician Aleksei Ukhtomsky (1875–1942). One of the greatest physiologists of the twentieth century, and one of the most tragic figures in Soviet science.

23. Director of the Institute of Organic Chemistry, academician Aleksandr Nesmeyanov (born 1899) claims to be the author of twelve hundred scientific works. This means that he has published an article or monograph every twelve days for more than forty years!

24. Nobel Prize winner and academician-chemist Nikolai Semenov (born 1896) is convinced that the leadership of the Academy of Sciences of the USSR cannot do without KGB members and other Communist Party bosses.

25. One of the most famous men in the USSR, and
founder of the now defunct science of agrobiology, academician Trofim Lysenko (1897–1976).

26. A member of the Academy of Medical Sciences of
the USSR, Oganes Baroyan is also director of the Scientific Research Institute of Microbiology, Immunology,
and Epidemiology, at Gamaleya, in Moscow. He likes to
remind his colleagues that he is also a KGB general.

27. Doctor of biology, and the director of the laboratory at Alma (Kazakhstan), Victor Inyushin (born 1938) believes that for political purposes, it is quite acceptable to use electronic instruments to manipulate the minds of the masses.

8. Biochemist Yuri Ovchinnikov (born 1934), vice-resident of the Academy of Sciences, has a special symathy for the Communist Party, and is the candidate ost likely to succeed to the presidency of the Academy.

29.   Olga Lepeshinskaya (1871–1963) was a member of the Academy of Medical Sciences, and a Bolshevik from Lenin's era. In 1950 she was awarded the Stalin Prize, first degree (200,000 rubles), for the fictitious invention of "living substance." For political reasons this "invention" has never been exposed in the Soviet Union.

30.   Doctor of biology Garri Abelev (born 1934) is known in scientific circles throughout the world for his research on cancer immunology, and for his unpublished article "Ethics as an Element of Organization of Science."

dates' degrees, but only one has the makings of a real neuropathologist."

In Karaganda, a regional capital in the Kazakh SSR, Tolem Ospanov, the head of the municipal health department, spoke to me along the same lines about the quality of local physicians and medical experts. Whereas Yuldashev in Uzbekistan was chiefly concerned with the health of the cattle breeders of Kyzyl Kum, Ospanov's responsibility was for the population of a large industrial town. But both of them were worried about the poor qualifications of local doctors and the ease with which they could obtain higher degrees. Ospanov deplored the fact that able and incompetent practitioners, the honest and the dishonest, were treated on an equal footing in the scientific and medical world of Kazakhstan. As far as the municipal service was concerned, he tried to make use of the ablest of whatever nationality. Six months after leaving Karaganda I heard that he had been dismissed from his post for "surrounding himself with Jews."

The anxiety of such men as Yuldashev, Ospanov, and Sofiev is by no means groundless, nor is the problem confined to medicine. In Siberia and Central Asia, on the Volga and in the Caucasus, Moldavia, and Belorussia—in all the outlying parts of our huge country the great mass of scientists is of third-rate caliber, even by Soviet standards. It resembles a tower of Babel with tens of thousands of candidates at the base, thousands of doctors halfway up, and hundreds of local academicians at the top. What do I mean by the comparison with Babel? Both the biblical monstrosity and the Soviet one are the work of human hands, intended as a symbol of human power and knowledge. God did not allow the first tower to be completed: "Go to, let us go down, and there confound their language, that they may not understand one another's speech. So the Lord scattered them abroad from thence upon the face of all the earth." As far as language goes, the Soviet situation is the opposite: members of many nations have a common language, that of science, in which they compose pseudoscientific monographs and lectures and concoct fraudulent dissertations. There is nothing surprising in this: crooks of all nations have always found a common language, thieves and extortionists need no interpreter. As for scattering the builders—the tower of pseudoscience in the eastern part of the USSR rises higher with every passing year, and I do not know what force can prevent it from reaching the heavens. . . .

The Mafia-like expansion of science in the borderlands was not invented by Turkmens and Uzbeks, however. Throughout the Union, scientists are spawned on a massive scale. In Belorussia, for instance, the number of people with higher degrees is increasing faster than in Central Asia—it doubled between 1965 and 1975—and the same is true in Moldavia. The fact that this means a deterioration of quality is acknowledged grudgingly by the press, but another aspect of the cancerous growth is never mentioned at all. At Kishenev and Tbilisi, Vilnius and Tallinn, Erevan and Tashkent, the value placed on a scientist's work depends entirely on his nationality. In Moldavia all important discoveries, without exception, have to be made by Moldavians, in Armenia by Armenians, in Georgia by Georgians. Here is a typical example.

Pavel Ch. is a student at Kishenev University with a passion for archaeology. For some years he has been taking part in local excavations conducted by a Moscow professor. He comes from a peasant family and his general education is rather sketchy, but he reads everything he can get hold of on the history of Moldavia and is determined to be an archaeologist. The professor encourages him and has even invited him to study for a higher degree in Moscow. All that is needed is that the director of the Historical Institute of the Moldavian Academy should give him a written recommendation, one of the innumerable documents that are issued all over the country for purposes of supervision and control. However, when the time came to draw up this document it turned out that Ch. was registered not as a Moldavian but as a Romanian. No matter that his forebears had lived for centuries on the banks of the Dniester: the Institute refused to issue a recommendation. The Moscow professor pleaded with the director, but the latter replied: "We want *Moldavian* scholars. The Romanian boyars oppressed the Moldavians for centuries, and therefore(!) Pavel shan't go to Moscow." The professor (G. B. Fedorov) was indignant. He himself had directed the first archaeological studies in Moldavia after the war, he had been working in that field for the past thirty years, and he was the best judge of who was fit to pursue higher studies in it. After a great deal of argument Pavel was allowed to go to Moscow and take a candidate's degree, but he was then relegated to an inferior job in the Institute, where he still works, simply because he was officially a Moldavian. Other pupils of Professor Fedorov's—Russians, Jews, Ukrainians— although born in the republic, have had to seek work elsewhere: Moldavia, it has been made clear to them, is for Moldavian scholars only.

In such circumstances the objective merits of a would-be scholar play no part whatsoever. To be accepted as a member of academic society in a minority republic, the first requisite is to show that one is a pure-blooded national. This double standard leads to absurd and sometimes tragic situations. In Tashkent I asked an elderly Tatar woman named Umidova, a corresponding member of the Academy of Sciences of the USSR, how she had managed to attain such a high position although she lived in Uzbekistan. Umidova, who is a competent physician and a popular lecturer, replied sadly: "I have had to pretend all my life to be an Uzbek."

The spirit of scientific nationalism rages unabashed in dozens and hundreds of research institutes from Kishenev to Erevan, from Dushanbe to Minsk. Year by year it becomes more aggressive and intolerant. Either by force or by subterfuge local scientists turn their institutes into a national preserve, using any and all methods to get rid of "foreigners."

In 1920 the University of Turkestan was solemnly opened at Tashkent, "by Lenin's personal order" as the phrase goes. As there were no qualified Uzbeks in those days, Moscow sent a large team of Russian scholars to Central Asia. One of the founders of the medical department (subsequently attached to the Tashkent Medical Institute) was Professor Valentin Feliksovich Voyno-Yasenetsky, a pioneer of the surgery of sepsis. Fifty years later the professor's granddaughter, herself a doctor, told me:

"The rector of the Institute said openly the other day that he did not mean to have a single European doctor or instructor in the place. Europeans (by which he meant chiefly Russians) were eligible as patients and nothing else. He had a list of fifty non-Uzbek instructors drawn up, and signed an order for their dismissal. The professors complained to Moscow; for a long time nothing happened, and then the rector was told in mild terms that his action was illegal and that the professors must be taken on again. Even then, with the support of the local Central Committee, the rector kept them unemployed for another two months."

This is not the only case of its kind in Tashkent, where it is almost impossible for non-Uzbeks with a higher education to get jobs. "Go back to Russia where you belong," they are told. Some of the fiercer nationalists resort to direct threats. One overpersistent Russian was told at the office of the Council of Ministers of Uzbekistan: "You get out of here double quick—if you make any more trouble we'll paint the roof with your blood." Even if this is put down to oriental fantasy, it is a curious way for an official to talk. At any rate, Moscow's sense of mission in Central Asia has not had exactly the results that the first generation of scholars would have expected. . . .

I did once meet a Russian in Tashkent who expressed complete confidence in the permanence of his job. Valery Sokolov, a power engineering specialist, agreed with me that "foreigners" were generally ousted from Uzbek institutions without scruple, but he reckoned his own job to be "safely in the hands of Russians and Jews." Valery is deputy head of the computer center; his boss, of course, is an Uzbek, but like other department heads he believes in having a Russian deputy. There is no danger in allowing an infidel to rise that far, since according to tradition he can rise no further: in a minority republic, a Russian second-in-command has reached the highest level to which he can aspire.

The intrigues and rivalries in university departments, laboratories, and research institutes in the national republics are only a surface agitation reflecting in a minor degree the profound national animosities that exist within the country as a whole. It is sufficient to spend a week in a national republic to realize that the vaunted "friendship of peoples" has collapsed like a house of cards. Perhaps the social and cultural atmosphere of Kazakhstan exemplifies this more clearly than anywhere else. It is a republic I have visited many times, and I have heard a good deal about conditions there from writer friends who live in Alma Ata, the capital.

The Kazakh SSR covers a million square miles, about four times the area of Texas. Out of its population of 14 million, only 4 million are Kazakhs. There are 6 million Russians, a million Ukrainians, a million Volga Germans who were resettled during the war, plus Tatars, Uzbeks, and others. The upsurge of local nationalism began in the late sixties, when leading officials were replaced and the Russian intelligentsia came under attack. The best actors and directors were dismissed from the excellent Russian theater; a student theatrical group, "Galerka," was broken up, as was a university society devoted to studying Russian literature of the 1920s. But the heaviest blow to cultural life consisted of a purge in the office of the literary journal *Prostor,* which for many years enjoyed a high reputation and was read regularly throughout the Soviet Union. Its editor, the writer Ivan Shukhov, published at his own risk forgotten works by Tsvetayeva and Mandelshtam, reminiscences of Andrei Platonov, and long or short stories by contemporary writers who could not get their works published in Moscow or Leningrad.[2]

At the end of the sixties *Prostor* was sharply criticized for "neglecting national themes"; shortly afterward it was placed under new editorship and obliged to publish, for the most part, translations from Kazakh. In the past, extracts from it used to appear in the London *Times* and *Le Monde;* now it sank to the level of a second-rate provincial journal. The publishing concern *Zhuzashi* was also reorganized, with the result that Russian writers in Kazakhstan could scarcely get any of their works accepted.

Among the leaders of the nationalist movement in literature and the arts are gifted young men such as the writer Anwar Alimzhanov, the poet Olzhas Suleymenov, and the scholar and publicist Murat Auezov. Given the strictness of the Soviet regime, there is no question of political or social reformism. Instead, the young intelligentsia have engaged in national mythology. The official myth in Alma Ata is that the Kazakhs, hitherto known to the world as a race of illiterate nomads, in fact had an important culture in bygone times. In the desert, it seems, are undiscovered cities of the ancestors of the present-day Kazakhs, containing evidence of literacy and high technical achievement. That being so, the Kazakhs are clearly more important than has been supposed. They do not need to be taught culture by anyone, but can themselves set an example in literature, science, and self-government. The nationalist doctrine even extends to the idea of a Kazakh mission to civilize the whole of Central Asia.

Nationalist scholars have more than once come to grief in the attempt to prove these strange propositions. One such incident got into the papers and gave joy to millions of readers of *Komsomolskaya Pravda.*[3] One day, a truck drove into the yard of the Institute of Philological Research at Alma Ata with a rock weighing eight tons, on which local linguists had found some mysterious inscriptions. A special epigraphic commission was set up under the director of the Institute, S. K. Kesenbayev of the Kazakh Academy of Sciences, assisted by two corresponding members of the Academy, R. G. Musabayev and A. T. Kaydarov. They studied the inscription and announced that it was an account of a hunting expedition by a prince named Bekar Tegin, some time between the sixth and fourth centuries B.C. It was thus clearly proved that the Kazakhs, twenty-five centuries ago, possessed a writing system and a state of their own. The Kazakh scholars were on the point of proclaiming the greatest discovery of the age, but fortunately the truth came to light before they had time to publish an imposing volume with the imprint of the Kazakh Academy. It turned out that the inscription had not been carved 2,500 years ago but in 1969, by a movie team who were shooting a film titled *Kyz-Zhibek,* and who copied the mysterious signs from a paleographic work by Academician S. E. Malov. They were in the language of the Orkhons, a Mongolian tribe; they had been deciphered long ago, and had nothing to do with Kazakhstan or any Prince Bekar Tegin. The translation ran in part: "We were worthless, and mistook small things for great."

This, one might have thought, would have put an end to the myth of Kazakh antiquity, but not at all. The scholars of Alma Ata are still searching for evidence of a lost Kazakh culture, and we may be confident that they will find it.

The most remarkable thing about the wave of Kazakh nationalism is that the dreams of the young intellectuals coincide with the official attitude. The young blurt out from time to time what their seniors more prudently conceal. The First Secretary of the party's Central Committee in Kazakhstan, a geologist named Din Muhamed Kunayev, is close to Suleymenov by race and mentality. They both understand that there cannot be a frontal attack against the Russians, but that they can be humiliated, exploited, and squeezed out by perfectly legal means and with the help of accepted slogans. Suleymenov and the other young Kazakhs know that they have become a power in the land, and take a dictatorial tone with their former Russian mentors.

Ivan Shchegolikhin, a Russian writer in Alma Ata, said to me: "When I meet Olzhas Suleymenov, whom I like and whom, so to speak, I introduced to literature, I am made to feel an intruder. He never stops reminding me that I and all the other non-Kazakhs who live here are undesirable aliens, invaders, and oppressors of his people."

Suleymenov himself said to another Russian writer: "Every one of your books must have a Kazakh hero, a 'positive' hero. You live here in Kazakhstan, and it is your business to write about it."

The Kazakh variety of nationalism is not the only one, but it is becoming more and more widespread. It is met with in western as well as eastern republics, especially Moldavia and Belorussia. As every historian knows, what is now the Moldavian SSR was for thousands of years a corridor through which people of Turkic or Iranian speech migrated from Asia into Europe. These nomadic herdsmen reached the mouth of the Danube and, being unable to cross the river with their cattle, settled on its banks. Greeks and Romans moved into the area from the southwest and Scandinavians from the north, followed by Slavs from the sixth century onward. And now, sweeping aside all this well-known history, the present Moldavian leaders require scholars to prove that the Moldavian people have been living between the Dniester and the Danube since the creation of the world, that they are of absolutely pure stock and have never mingled with any other race. Those who refused to comply with the directive have been ousted from the Moldavian Academy, while the remaining historians and archaeologists are forced to devote themselves to spurious lines of study. Party pressure is already responsible for countless forgeries and falsifications in museums, textbooks, and historical monographs. The effect of this historical barrage is simply to justify the dismissal of "foreigners" and their exclusion from the political, social, and academic life of the republic.

In Georgia and Armenia the situation is different. There the young nationalist intelligentsia has no need to rewrite history or to concoct it out of nothing. The ancient culture of these once independent lands has no need of boosting or embellishment. But Georgian and Armenian scholars have also come to adopt an arrogant tone with their Russian colleagues. Dr. V., a Moscow specialist in radioelectronics, was a member of a large commission sent to the Georgian Institute of Cybernetics at Tbilisi to take delivery of some scientific material which the Georgians had contracted to supply for military purposes. The director of the Institute, Academician Chavchanidze, made it clear that the work had not been done properly but concluded his report, smiling ironically, with the words: "We are confident you will accept our study with gratitude." Outside the meeting V. complained to the leader of the commission, a senior military engineer, that the Georgian scientists were trying to make fools of them and that the Army should not pay for a worthless report. "Unfortunately," was the reply, "we must pay for it and even say 'thank you' to that swindler Chavchanidze. The orders are that we must encourage science in the national republics."

If the republic authorities take this tone with specialists from Moscow, it is not hard to imagine how they behave to Russian scholars and others who live permanently in their midst. The pressure on the nonlocal intelligentsia has been so intensified that writers, journalists, doctors, and scholars are beginning to withdraw to Russia proper. There is as yet no mass exodus, but thousands are feeling the urge to escape from cities in the national republics. When in the USSR I received letters from many friends who had lived for decades in Uzbekistan, Tadzhikistan, the Ukraine, or the Caucasus, asking me to help them find a place to live in any Central Russian town, the atmosphere of threats and pressure was more than many of them could stand.

The Moscow authorities of course know about the growth of local nationalism, and are keeping a close watch on what may be the gravest of all dangers for them. The top party leaders, however, do not venture to take any decisive steps.[4] But even apart from the general political issue, what is going on in the republics is of major importance from the point of view of "manipulated science." The growth of a nationalist consciousness implies a degree of mental independence, a possibility of centrifugal trends. Even such peaceful nationalism as that of Armenia or Belorussia betokens a dangerous immunity to control. So far, developments in republican academies and institutes are only directed against "outsiders" and do not threaten the machinery of the state, and the central authorities pretend that all is in order. It is not considered good form to talk publicly about nationalism, even the scientific variety. Newspaper articles and speeches still refer to the "granite unity" of the Soviet state and its "unshakable monolithic power"; but it is known that some precautions are being taken at top level.

For instance, in systems of nationwide importance such as railroads, civil aviation, border control, major telephone and telegraph connections, etc., preference is given to employing Russians. This applies even more clearly to establishments connected with the space program. In Tashkent I asked a Russian woman, a scientific officer at the Meteorological Institute concerned with the launching of space satellites, what the proportion was of Russian to Uzbek employees. She replied that there were no Uzbeks in the Institute at all. "We do real work there," she added vehemently, "and we need real workers, not those idlers and scroungers."

The upsurge of nationalist passions and reciprocal chauvinism in the borderlands is reminiscent of what happens when you put a saucepan of milk on the fire. First a few bubbles appear around the edge, then more, until the milk is seething at every point of contact with the hot metal. Then the bubbles spread swiftly to the center and—well, every housewife knows that it is time to whisk the saucepan off the fire.

Historical predictions are risky and I will not venture to prophesy the outcome of nationalism in Russia, even of scientific nationalism. I will, however, quote the opinion of a physicist and academician, Mikhail Aleksandrovich Leontovich.[5] I once asked him as we were traveling in the subway what he thought of national conflicts in Soviet science. His childlike blue eyes gleamed as he leaned toward me and whispered, so as not to shock the other passengers: "Nationalism means that your own shit smells good, and chauvinism means that *only* shit smells good."

The Christian philosopher Nikolai Berdyaev, in his book *The Meaning of History,* argued that the Jews, unlike other peoples, were put into the world for a special metaphysical purpose. Like yeast in unleavened dough, their function was to ensure that the spiritual and creative principle never died out of human affairs. Although myself a Jew, I do not feel capable of either confirming or refuting this supposition. The divine purpose is likely to remain inscrutable. Leaving aside such higher thoughts, I shall consider simply the question of the part played by Jews in Soviet science. The subject is of importance because the relationship of Jews to scientific and educational matters distinguishes them sharply from the other peoples of the Soviet Union.

Here are some figures.

Soviet Jews number two million, i.e. less than 1 percent of the country's total population. As against this, they represent 5 percent of people with higher education and about 10 percent of scientists. As is known, in Tsarist Russia Jewish educational opportunities were limited. The maximum ratio of 3 percent in gymnasiums (high schools) and Imperial universities meant that numbers of young men and women left their country to study in Europe or America. At the beginning of the twentieth century the desire for education took a primary place in the aspirations of Russian Jews. When the February Revolution of 1917 opened the doors of schools and universities to all nationalities in Russia, the Jews were best placed psychologically to take advantage of the new order. Thousands of young people from townships in the Pale of Settlement flocked to the university cities. Convinced that learning would free them once and for all from their subordinate position, they renounced everything they had previously worshiped, including their ancestral religion and sense of national identity.

For twenty years after the Revolution they were buoyed up by hope. From an ethnic group they became a social one distinguished mainly by its high educational level. Today practically all adult Jews in the USSR have completed secondary school, and 2 percent have received higher education.[6] In many fields of science —physics, especially nuclear physics, and some branches of chemistry—most of the leading specialists are Jews. In 1973, 10 percent of members and corresponding members of the Academy of Sciences of the USSR were Jews; in 1946 the figure was 14 percent. After World War II, however, the state authorities openly adopted a policy of excluding Jews from academic life. This policy has been more severe at some times than at others, but essentially it has remained unchanged for thirty years. In the 1970s, after Jews began to emigrate, it entered a phase of special virulence.

Unlike the position in Tsarist times, there is no legislative basis for Soviet measures which prevent Jews from entering universities, institutes, and even secondary schools of languages and mathematics. It is known, however, that there are many places of higher learning to which Jews are not admitted at all, particularly military and party schools and those connected with diplomacy and foreign trade. It is also very difficult for a young man of Jewish extraction to enroll in a university faculty of journalism, history, philosophy, or philology. Several universities, especially those of Kiev, Leningrad, and Moscow, are noted for their especially hostile attitude to Jewish enrollment. Many schools of technology are also hard for Jews to get into. Facts of this sort are generally concealed, but here are some figures. In 1971 the chemical faculty of the Kiev Polytechnic Institute received 498 applications for admission, as follows:

	UKRAINIANS	RUSSIANS	JEWS
Applications	356	127	15
Passed examination	202	96	2
Admitted	148	56	1

In other words, nearly half the Ukrainians who applied were admitted, but only one fifteenth of the Jews. I showed these figures to a large group of scientists at Kiev. Their reactions varied considerably—some were anti-Semitic—but they agreed that the figures were typical of all Ukrainian institutes, and that the ratios had not changed much in the last fifteen or twenty years. No Jews at all are admitted to Kiev University. My Kiev informants even assured me that there was a secret university regulation excluding people with a "capitalist fatherland." I doubt that such a document really exists, however: rectors and prorectors have been imbued with such a keen sense of what is expected of them that there is no need for party bosses to put anything so compromising on paper. Altogether, the Soviet authorities have managed quietly to reduce the Tsarist quota of 3 percent at universities, and many institutes, to 0.3 percent. Jews get most of their higher education at night schools and from correspondence courses. At Moscow University the position for some years past has been that among 150 students attending the daytime course of mechanics and mathematics there are not more than five or six Jews, whereas at the evening course there are several dozen of them. The Jews are also disadvantaged in that they are not admitted to postgraduate studies, so that the broad path to higher academic attainment is largely closed to them.

Let us take the position of a graduate from a university or technical institute. As a Jew, he will find it hard to get a job at a research institute, even as a laboratory assistant. Although the rule is that the university will see to it that graduates of daytime courses find jobs in their special sphere, in practice the institutes find various pretexts for turning down applicants of unsuitable nationality (Point 5 in the regulation questionnaire). What will happen to the Jewish graduate depends largely on his own resource. I have known young physicists, chemists, and mathematicians, anxious at all costs to pursue a scientific career, who had graduated from technical institutes and even presented doctoral theses, but still had to apply to thirty or forty research institutes before they found a job. They had interesting tales to tell of their wanderings. A mathematician who applied to the Automatic Control System Institute of the Moscow municipality was asked to fill out a form in two copies, on one of which, to be kept personally by the director, he undertook not to settle in any foreign country. Thus the director assured himself of an alibi even before hiring the applicant. Another Jewish mathematician spent a year filling out forms and attending interviews with senior members of the Institute of Economic Research under the State Planning Commission of the USSR. The director had apparently made up his mind to engage the young man when he suddenly broke off contact and sent a message through his secretary that he was not hiring any more Jews. It turned out that he had received a warning from higher authority that a commission might be appointed to look into his staffing policy. At this he had become alarmed—it is no light matter to be accused of "pro-Jewish inclinations."

One more story. A Jewish chemist, a candidate of sciences whose particular specialty was in great demand, visited several institutes in search of work. All his interviews with the deputy directors began and ended in the same way: "Yes, yes. We do need you, very badly, and I'll try and get the director to take you on. But—well, you know how it is, there's a quota." The deputy would go off to see the director, and return looking gloomy. "Yes, it's too bad, but the quota for Jews is already filled."

As a rule, however, a young Jewish specialist who applies to a research institute doesn't get to see either the director or his deputy, but is interviewed by a laboratory head. Many people think the latter has special instructions to refuse Jewish applicants, but in fact he doesn't need any. He well knows that if he engages a Jew and if, later on, the latter tries to emigrate to Israel, he himself will be dismissed on account of "failure to exercise proper moral control." From the director's point of view, a laboratory chief who takes on a Jewish assistant is in the position of a special kind of hostage. This happened to one Akishin, a strong, energetic man with a good party and social record, a section head in the

Nuclear Physics Research Institute at Moscow University, who was rising rapidly in his career. He got some Jews on his staff to write a doctoral thesis for him, but paid dearly for it when they manifested a desire to go to Israel. Akishin was investigated, judged, and condemned in short order. The most unexpected charges were made against him. For instance, when Todor Zhivkov, the Bulgarian communist leader, had come to confer a decoration on the university, the members of Akishin's section had not applauded very loudly—was that not strange? Or again, when the university staff had assembled on Leninsky Prospekt in their capacity of "popular masses" to welcome some newly arrived foreign guest, the full complement of Akishin's section had not turned up at the appointed place. A few more charges like this were hurled at him, and good-bye to his position in the Institute— he would be lucky to hold on to his party card. . . . It need hardly be said that Akishin, with his career in ruins, will never again hire a Jew.

In the same kind of way, heads of sections and laboratory chiefs are trained to prevent Jews from presenting doctoral theses. If an ordinary physicist or mathematician emigrates to Israel, it is less serious for his boss than if he is a man with a higher degree. A candidate is bad enough, but if it is a doctor, the whole Institute is convulsed, and if it is a corresponding member, the Academy is shaken from end to end. Thanks to all this buffeting, and the certainty of trouble in both professional and party circles, heads of departments do not need special orders to persuade them of the simple truth that it is a mistake to allow a Jew to take a doctor's degree. Everyone reacts to this situation in his own way. The Higher Examinations Board does not hesitate to express its disapproval when an Academic Council recommends a Jew for the doctorate, and it will allow his thesis to stew for two or three years. The Academic Councils have taken the hint and find all sorts of reasons for not considering theses by those stigmatized under Point 5. If a Council is obdurate, it will itself be disbanded, as was the Academic Council of the Research Institute of Radiophysics at Gorky, as a penalty for encouraging "undesirable elements."

Notwithstanding all this, many Jews work in Soviet research institutes, their theses occasionally get through the Higher Examinations Board, and some even become corresponding members of the Academy. This is not by accident or negligence, but is also part of a pattern. The anti-Semitism of the Soviet state is inconsistent as a matter of principle. Two directly opposite tendencies are at work. On the one hand, district and municipal party committees keep a close watch on the national, or rather the Jewish, aspect of the composition of scientific cadres; so do personnel departments and the KGB units that supervise them. Directors and Academic Councils of research institutes take care that a due proportion is observed. Secret institutes ceased to take on Jewish staff a long time ago; academic publishing houses have secret instructions not to publish too many books by Jewish authors.[7] On the other hand Soviet propaganda in the West insists on how well Jews are treated in the USSR, and gives figures of the large number of them engaged in scientific work.[8]

In 1968 and 1970 the following Jews were elected corresponding members of the Academy of Sciences of the USSR: F. L. Shapiro (nuclear physics), I. I. Gurevich (nuclear physics), Yu. M. Kagan (nuclear physics), Yu. M. Krasny (physics), P. V. Geld (chemistry), E. S. Fradkin (nuclear physics), and Ya. Sh. Shur (physics). But not long afterward, at the end of 1973 or beginning of 1974, academic circles in Moscow heard that the president of the Academy, M. V. Keldysh, had been asked by the Politburo how many years it would take to eliminate the Jewish element from the higher reaches of Soviet science. Keldysh's estimate was seven years. A few months later, in summer 1974, there was another change of course: Yu. A. Ovchinnikov, vice-president of the Academy, in a conversation with biologists about the need for secret experiments to increase military potential, said that in order to speed up this work the Academy was willing for any number of Jews to be employed in a special laboratory.[9]

In this way Soviet policy, and therefore the Academy, lurches from side to side. As the saying goes, "It's too hot to hold, but he can't bear to throw it away"; or, as a Moscow poet wrote: "Wherever history is being fearlessly patched up and cut about, you'll find Jews as scapegoats who are also milch cows." The leaders, in short, would like to purge Soviet science of Jews altogether, but they realize this would do them no good from either the political or the military point of view. They need the milch cow, and so the problem remains in an indeterminate state.

In autumn 1975, when Gury Ivanovich Marchuk took over from M. M. Lavrentyev as head of the Siberian department of the

Academy of Sciences of the USSR, he visited the institutions under his authority and laid down, among other things, that no Jews were to be dismissed but no new ones were to be taken on. This is the latest expression of political wisdom as regards Soviet citizens of Jewish race. Specialists going about their work conscientiously may be harassed in small matters, but they are allowed to keep their jobs; the next generation, however, will find the doors of research institutes firmly closed against them.

Propagandists in closed party sessions try to justify this policy on the ground that Jews are suspect since the migration to Israel began. But the drive to exclude Jews from science and culture began in Stalin's day, a quarter of a century before the exodus. Ever since then it has presented the leaders with the vexing problem of how to avoid forfeiting Jewish skills altogether, especially in the armaments sphere. The formula now is no different from what it was in Stalin's day: "Better not have any Jews, but if you need this particular one . . ." The same, of course, applied in Nazi Germany, where there are said to have been about 30,000 "useful Jews."

On the one hand it is impossible to eliminate the Jews from Soviet science, but on the other, anti-Semitism is a deeply implanted tradition of the Soviet state. How do matters look today from the point of view of the million-strong scientific army?

In the first place, there are Jews at the head of several of the most important research institutes under the Academy of Sciences. All of them are completely dependent on the regime and therefore absolutely loyal to it. They, and a few hundred more heads of laboratories, are the chief purveyors of ideas, inventions, and discoveries, especially for military purposes.[10] These top scientists are well looked after, but there are a large number of ordinary specialists whose condition is very different. Many are highly skilled, and all are anxious to serve the cause of science and the Soviet economy, but their ambitions are frustrated by their Jewish origin. By and large, nonmilitary discoveries do not interest anyone in Russia, and if they are made by Jews they are positively unpopular. Scientists of a passive turn of mind resign themselves to this fact, but others pester the Central Committee with their projects, write to the newspapers, and address moving appeals to ministries. They long to enrich the fatherland with their intellectual gifts, which are indeed considerable, and they cannot understand why the fatherland is so blind to its own economic interests.

This situation has been going on for years and decades: the Soviet state absentmindedly cuffs its unruly children from time to time and nips all their good intentions in the bud, while they continue naïvely to explain at every opportunity how useful their ideas, inventions, and discoveries might be to the country. The only aspect which puzzled me for some time was why the state officials and academic administrators, in addition to stifling Jewish initiative, showed such irritation and even malevolence toward their victims. This attitude is constantly observable in ministries, newspaper offices, and the management of many research institutes. It is, of course, a general psychological principle that we hate those to whom we do evil, but there is also a specific reason in the field of science. The scientists' creative urge and their keen desire to see their ideas put to practical use highlight the indolence and incompetence of the bureaucracy and the economic absurdity of the whole system. Not surprisingly, officialdom regards these pushy scientists as a thorn in the flesh. In a society where goods are distributed strictly in accordance with rank, where commendation or punishment descends on the citizen from above and depends entirely on his superiors, people who wish to achieve success by their own efforts are a discordant element. Here are Jewish scientists, inventors, and organizers, all with higher degrees and outstanding skills, hoping to obtain social recognition by their own merits instead of by the grace of their appointed rulers. What could be more impudent and more disturbing?

The insidious attempt of talent to assert itself from below is an attack on the very foundations of Soviet power. Jewish initiative and energy are an unmistakable reminder to the country's rulers that there is a more dynamic society which, despite predictions, not only is not withering away but, in the seventh decade of Soviet power, is still supplying the country with food and machinery. Not only do Jewish specialists fail to fit tidily into the pattern of manipulated science, but their very existence is contrary to the forms of Soviet social consciousness. "They're too smart altogether!"—as a senior official of a committee on inventions and rationalization put it in a private conversation. It doesn't pay to be smart in the Soviet Union, however right and however talented you may be.

In recent years the dislike of "smartness" has been given a more philosophical basis. At the official and semiofficial level there has been serious consideration of a plan to make the composition of the Soviet intelligentsia by nationalities proportionate to the population ratio of the nationalities themselves. The larger a nationality is, the more intellectuals it would be allowed to have. The Jews, being less than 1 percent of the country's population, would clearly not be entitled to a large intelligentsia, and Jewish admissions to places of higher learning would be restricted until the number was suitably cut down. Obviously this enforced "equality" would be no more or less than a degradation of the Jews, forcing them out of areas of science and culture in which their knowledge and experience have given them the lead for decades. But while social engineers debate such projects, many employment boards have already put them into practical application.

Such is the outline of the basic conflict between the Kremlin and the Jewish intelligentsia, scientific or otherwise. To some extent it has been alleviated by the flow of emigration. This is diverse in character, but the great majority of Jews who leave the Soviet Union do so for the same reason as an earlier generation left Tsarist Russia, namely that harsh government control in the educational and cultural field prevents able individuals from taking their due place in society.

Emigration, of course, will not and cannot solve the general problem of Jews in the Soviet Union. However many doctors and candidates of science leave the country, many more will remain, and those who do are confronted with grave threats to their moral integrity. Many people who show great determination in pursuing their investigations or qualifying for a higher degree find their strength of character undermined by the routine of a scientific institute. A Jewish researcher who is constantly being reminded that he is a second-class citizen is oppressed by a sense of doom and failure. The creative urge soon becomes atrophied; the researcher goes on with his work as a matter of routine and because he needs to make a living. I have noticed this phenomenon especially in research officers over the age of forty, working in secret establishments.

The unnatural atmosphere in which Jewish scientists are obliged to live also gives birth to a more cynical type. "If you want to get on, you have to make up to people"—this typically Soviet saying is often heard from Jews who have attained a certain position in the scientific world. So a researcher, with fear in his heart and an obsequious smile on his face, will "make up" to his chief or curry favor at party meetings, in the belief that only such servility will buy him the right to work as he sees fit. How often does one see overpliant young Jewish people of this kind—how little they do for science, and how little enjoyment they get from it! They are popular with the top brass, however, because they can be used to play dirty tricks on others, and, if need be, they can be thrown out without compunction—after all, they're only Jews. . . .

A few Jewish scientists manage to crawl through the dirt and humiliation to high positions—and these should be given a wide berth. Not for nothing did the ancient Romans say that freedmen were the worst of mankind. The image of a liberated slave armed with the foreman's whip comes before my eyes when I recall the havoc among Soviet geneticists in the late 1940s. At the session of the Lenin Academy of Agricultural Sciences in August 1948, when Lysenko, armed with dictatorial power, mercilessly attacked biologists and the basic principles of their science, a surprising number of his entourage were Jews: Prezent, Belenky, Rubin, Kuperman, Kushner, Feyginson, and Khalifman were particularly harsh and aggressive. The country at large was swept by the wave of anti-Semitism unleashed by Stalin—the Jewish anti-Fascist committee had been dissolved and its members shot, the principal Jewish writers and artists had been hounded to death—but Isay Prezent, Neo Belenky, Boris Rubin, and others were in Lysenko's camp, dealing despotically with his enemies and their own. Not long before the alleged "doctors' plot" to murder Stalin, the same Boris Rubin was put in charge of the department of physiology at Moscow University, a hotbed of anti-Semitism, and he occupies that position to this day. Years later, when Lysenko's dictatorship was a thing of the past, I asked Rubin why he had been so aggressive at the 1948 session. He replied, as though it were perfectly natural: "There was nothing else to do, you either ate or got eaten. Remember what happened to Rapoport."

I knew Professor Rapoport, an eminent geneticist who was driven from the field of biology after the 1948 session, and I shall tell his story presently. As for the formula "There was nothing else to do," this is still as lively as ever. In the summer of 1973, having joined with other academicians in signing an open letter against A. D. Sakharov, Aleksandr Naumovich Frumkin, director of the Academy Institute of Electrochemistry, was entertaining a group of close friends at his country villa. A gloomy, taciturn man seventy-eight years of age, Frumkin declared without false shame that if the academicians had not drawn up and signed the disgraceful letter attacking Sakharov, the whole Academy would have been in serious trouble. This had been made clear to Keldysh at a session of the Central Committee, which had all kinds of punitive measures in store if the Academy should prove recalcitrant. So the act of signing was, first and foremost, a means of preserving the chief center of Russian science and scholarship.

After Frumkin's death a relative of his, a professor of chemistry, gave me a still clearer idea of the old man's position. Frumkin never forgot that he was a Jew. Despite the rewards and dignities showered on him he knew that the slightest mistake might be fatal to his position as Institute director. If the Academy came under fire, he well knew that the first to suffer would be himself and others like him. He did not say to his close friends, "There was nothing else to do," but his conduct was summed up in that saving formula. All his public acts—like those of Khariton, Braunshtein, Frank, or Vul—were dictated by the sense of insecurity by which all Jewish academicians are affected, consciously or otherwise. The demoralization intensifies as it spreads downward, for the position of a Jewish laboratory head is even more precarious.

It is curious that demoralization should be linked with nationality, yet my experience is that in present-day Russia national feeling is one of the chief factors threatening morale. We have seen how inflated political nationalism has encouraged empty scientific pretensions on the part of young and not so young people in Turkestan, Azerbaijan, and Kirgiz. Special rights conferred on a nationality basis generally breed corruption. While overweening nationalism prevails in the minority republics, most Jewish scientists in Russia would be glad to forget that they belong to a different race. They are not allowed to forget it, and this too has a corrupting effect on the weaker of them. Thus the overvaluing of national characteristics in Central Asia has a similar effect to the arbitrary depreciation of the mental and creative powers of Jewish intellectuals.

The inferiority complex from which Soviet Jews are made to suffer sometimes has tragicomic effects. I have already described my talks with the academician and chemist Solomon Aaronovich Giller of Riga (1915–1974). Giller was the director of a research institute of international repute, yet even in his own apartment he never spoke except in a low voice. If the conversation turned to politics or national scientific cadres, a delicate subject in Latvia, he would drop his voice to a whisper, as his frightened gaze roamed over the familiar walls in search of a hidden listening apparatus. Authority of every kind—Moscow or Riga, party or academic—filled him with terror, and for fear of unpleasant surprises he would avoid meeting strangers for days on end. At those times his secretary at the Institute told callers that he was away, and his wife did the same at home. Above all he was terrified of hiring Jews. "If I take on a no-good Latvian I am complimented, but if I were to take Jews it would be the ruin of me and of the Institute," he whispered, at the same time begging me not to tell anyone that there were three Jews in the Riga Institute of Organic Synthesis.

Unlike Giller, the Moscow immunologist Georgi Yakovlevich Svet-Moldavsky (born 1926) could not be called a coward by anyone. In addition to his brilliant work on the immunology of tumors, Professor Svet-Moldavsky has for a year or more been fighting an incurable disease of his own. His staff at the Oncology Institute of the All-Union Academy of Medical Sciences regard him as a man of exceptional courage; so does a friend of mine who recently made a documentary film about him and allowed me to hear recordings of his lectures, interviews, and talks with his staff. In conversations lasting many hours, only some of which were used for the film, he came across as an outstanding character. At one point, however, at the beginning of a new tape, a short dialogue revealed a corner of his personality that he was clearly not in the habit of disclosing. He was telling my director friend about the problems of staffing a large institute, and how for many years he had not been able to hire a doctor named Misha Vladimirsky. Misha had written a brilliant thesis and was a first-class experimenter, but he was still working in the casualty department, as they could not get him into the Institute proper.

"Why?" said the film director. "Anything wrong with his personal details?"

"Well, yes, he's a Jew."

"But what about yourself, does it make a difference in your case?"

"Myself?" the professor replied in a low, toneless voice. "Myself . . . not really . . . Well, of course it does . . . How should I put it? . . . This is off the record . . ." A pause, and a click as the recorder was switched off. It was not difficult to guess that the interviewer had incautiously touched a sore spot.

Such is the atmosphere in which a Jewish scientist in his own home country begins and ends his working day: ethical compromises, stresses, and strains which some find natural and others conceal as best they can, but which finally corrupt all alike. It is the more difficult to resist because official "planned" anti-Semitism induces waves of anti-Jewish feeling in the "private sector" also. Many institute directors and laboratory heads now have a kind of vested interest of their own in persecuting Jews. What is more, the harassed victims end by adopting counterprejudices of their own—a national antipathy and even hatred of everything Russian. I have even heard scholars say that anti-Semitism is natural to every Russian, and especially to every educated Russian. This, however, I do not believe. In the forties, fifties, and seventies I have known many Russian scientists to whom official anti-Semitism was no less repellent than to their Jewish colleagues. Some academic administrators defended Jewish staff at the risk of their own position. A passage deleted by the censor from my book *On the Track of Those Retreating* (Moscow, 1963) described a significant occurrence in the career of the epidemiologist Nikolai Ivanovich Khodukin.

In winter 1952–1953 Khodukin, then director of the Tashkent Institute of Vaccines and Serums, was told orally by the district party committee to get rid of all Jews on his staff. On the same day he assembled those concerned in his office and spoke on these lines: "There have been many dark periods in the history of our country. They passed, and so will this. Do not submit your resignations. If they throw you out, I will resign myself."

Khodukin's remarks soon became known to the municipal party authorities, and it was openly said in Tashkent that his days were numbered. He would probably be arrested, or, at best, expelled from the party, stripped of his doctor's degree, and sent to work in some obscure village. All the signs were indeed to this effect. The District Committee got to work preparing a case against the Institute and its director. The witch-hunters were slightly embarrassed by the fact that the "protector of Zionists" turned out to be an old party member, the son of a railroad worker, and a veteran of the Civil War, not to speak of having freed Uzbekistan from the scourge of malaria; but in February 1953 such details did not count for much. A general meeting of the Institute, at which Khodukin was to be dismissed, was set for the beginning of March. But then Stalin died, the party members fell into confusion, and the case remained in suspense. A few weeks later Khodukin came into the Institute and found most of the staff in the lobby discussing a broadcast they had just heard—the Jewish doctors' plot to murder Stalin was, it appeared, a mare's nest. . . .

"God be praised!" Khodukin exclaimed. "At last I can breathe freely, and at last I need not be ashamed to call myself a Russian professor."

This story was told me at Tashkent by two former members of Khodukin's staff, Evgeniya Yakovlevna Shterngold, doctor of medical sciences, and Maria Zalmonovna Leytman, candidate of science. They added: "Nikolai Ivanovich was a man you could trust absolutely. We were quite certain that if we left the Institute he would go too." When I discussed the episode with Jewish friends in Moscow, they said: "Yes, very touching, but you must admit we have a harder time here than *they* do." Whether Russians or Jews have a harder time in Russia, I don't know; but I think those who suffer most, regardless of nationality, are those who still have a sensitive conscience.

There is one more story I promised to tell: that of Yosif Abramovich Rapoport.

In the course of thirty years I met him not more than three or four times, but each meeting was remarkable in one way or another. The first was in August 1948 at the famous session of the Academy of Agricultural Sciences which destroyed the development of biology in the Soviet Union for the next ten years. At that time, however, as a budding journalist reporting for the Moscow *Bolshevik,* I had no idea of the special importance of the occasion. My assignment was to go to the Ministry of Agriculture club on Orlikovsky Perevlok and write about the achievements of Soviet science. Scientific victories, my editor impressed on me, were a proof of the superiority of socialism, and it was my business to find them. If a particular research institute had no victories to its credit, the reporter should leave it alone and discover some more typical and progressive scientists. The editor did not want to hear of anything except the "resounding victories of Soviet Science."

Fortunately I did not record any of the "achievements" on that occasion, as the events which began on the second day of the Academy session were far from "typical" and would certainly not have been published if I had written about them. I therefore ceased to record anything, and only looked on and listened with growing astonishment. The proceedings have all been described since then: how newly appointed academicians loudly booed and whistled at the veteran professors who opposed Lysenko; how M. M. Zavadovsky collapsed with a heart attack in the middle of his lecture; and how, at the last moment and in panic haste, Academician Zhukovsky and Professors Polyakov and Alikhanian mounted the rostrum to do penance for the error of their ways. Amid the kaleidoscopic picture of those tragic and comic events I can still see the figure of Yosif Rapoport, with his black curls and boyish expression. He looked very handsome in his military tunic without shoulder boards but with rows of medal ribbons on his chest. Even the black bandage over an empty eye socket did not disfigure him, but lent a keener expression to his pale nervous face. His address, too—he was a candidate of science at the time—was nervous but firm in tone. He confined himself in the main to purely scientific points: that genes were a physical reality, that mutations could be controlled, that genetics had been of great benefit to mankind and could produce many more useful results. He ended with the blameless statement that "only on the basis of honesty and criticism of our own mistakes can we advance toward the great successes to which our country calls us." The verbatim record[11] states that Y. A. Rapoport of the Institute of Cytology, Histology, and Embryology of the Academy of Sciences of the USSR was "faintly applauded" on leaving the rostrum. As I recall, the applause was mingled with hostile whistling.

The next speaker devoted himself to an attack on Rapoport, who replied in kind with rapierlike interjections from the hall. Next day it became known that Comrade Stalin himself had studied Lysenko's address and approved it. From that moment on, everyone who had ventured to oppose Lysenko hastened to do penance and beg forgiveness. "A sleepless night helped me to think over my conduct," whined Academician P. M. Zhukovsky, a pupil and assistant of Nikolai Vavilov, adding vehemently that "we are not using the press as we should to combat foreign reactionaries in the field of biology. . . . I shall carry on this fight, and I attach political importance to it."[12] The geneticist S. A. Alikhanian went further still: "As from tomorrow I shall not only start to purge all my scientific activity of old reactionary Weismannist-Morganist views, but I shall make all my pupils and comrades do likewise." Rapoport spoke again, not to pour contempt on himself and his teachers, but to repeat: "I believe in the authenticity of a properly conducted experiment, I believe in the existence of genes and in the future of genetics." Strange to say, these final remarks of his were not included in the verbatim report.

In the corridors, Rapoport was the object of commiserating comment. Colleagues recalled that he had spent the war in the front line and had also belonged to a partisan unit. He had been recommended for the distinction of a Hero of the Soviet Union, but had not been awarded it because he was a Jew. The general conviction was that he would now pay for his obstinacy, and was certain to be packed off to a labor camp.

Rapoport was not arrested, but suffered every other penalty to which a recalcitrant Soviet scholar is liable. He was expelled from the party, dismissed from his Institute, and stripped of his degree. He could get no further employment as a biologist, and switched to geology, where he soon propounded a new method of determining the age of coal strata from analysis of the pollen of fossil plants. After Stalin's death he was able to qualify as a candidate of geological sciences.

Ten years after the cataclysm of 1948, the first genetics conference was held in a lecture hall at Moscow University. There was a huge attendance; people embraced one another in the corridors, laughing happily and even brushing away tears. At last they could once more devote themselves without fear to a science which for years they had been forced to brand as an outcropping of religious superstition and a bait of the imperialist powers. Rapoport took the floor and was greeted with an ovation; the news had gone around that Academician Nikolai Semenov had invited him to work at the Institute of Physical Chemistry. There he had already begun to put into practice the ideas he had proclaimed at the 1948 session. He produced and tested chemical substances known as mutagenes, which made it possible to alter the heredity of cultivated plants. Several new varieties had already been bred by this means, chiefly, it is true, in foreign countries. Like others, I was delighted with the successes of this remarkable man. He was still handsome, though the once abundant hair over his high forehead had thinned considerably and his shoulders were rounded. I sought him out in a lecture room, introduced myself, and asked if I might come to see him, as I was much interested in the problem of mutagenesis. He looked at me in silence for a moment, and then asked: "What did you say your name was? You've written about genetics—and what about the book on Lysenko?" The work on Lysenko was in fact by my father, Aleksandr Popovsky. It had been reprinted ten times during the Lysenko dictatorship and translated into several languages, but I myself had nothing to do with it. I tried to explain this to Rapoport, but he did not give me time. His brows beetled and his face flushed with anger as he pointed to the door and cried: "Get out this minute!" I looked around at the bystanders, but met only accusing glances. No one wanted to hear my explanations. For the previous ten years journalists and writers had concocted so many fables about the Morganists and Mendelians that geneticists no longer trusted them an inch; and they were right. Hanging my head, I left the Congress.

Another ten years passed, and more. I had written one book about Nikolai Vavilov, and a second which was refused for publication. I thought about Rapoport more and more—what a subject *he* would be for a book, yet how dare I approach him? In winter 1971 a female friend of mine telephoned from Moscow Radio. "We're recording a talk by Rapoport. I spoke to him about you. He's read your recent books and is prepared to see you. Come quickly." I hurried to the studio, and the three of us sat and talked late in the evening, after the recording session was over. Rapoport was now stooped and gray, but gave no impression of being aged. He wore a faded, unbuttoned shirt with no tie, and a crumpled suit of nondescript color: not that he was a poor man at the time, but for years he had had to count every kopeck, and the habit had stuck. He suffered from asthma, and took a sip of medicine every few minutes. But he enjoyed the conversation. A book about Vavilov—splendid! Vavilov was his ideal. But there was no reason to write about himself, Rapoport, except maybe after his death, which wouldn't be long now. I asked if I could visit him at home. "Certainly," he said with a smile, "but I warn you: I can stand up for myself."

Driving about the city on the way home, Rapoport grew more expansive. I questioned him about the Nobel prize. Yes, he said, in 1957 and again in 1964 the Nobel committee in Stockholm had asked the Academy of Sciences its view on the idea of awarding the prize to the creator of chemical mutagenesis. The first inquiry came to Rapoport's attention months later, from a chance informant; the Academy had turned it down on some specious pretext, but it was reliably thought that the real reason was his Jewish nationality. On the second occasion the Presidium agreed on condition that Rapoport first become a party member. Having joined the party in wartime and been expelled from it at Lysenko's bidding, Yosif Abramovich was not disposed to fall in with the Academy's wishes. . . .

I did not succeed in writing a book about Rapoport. Modest and ascetic, he was certainly able to "stand up for himself," and one of his tenets was that there was a sharp difference between great men of science, who should be extolled and written about, and "draft horses" like himself, who had no claim to fame and whose place was between the shafts as long as they were capable of working.

This was indeed a different view from that of the "useful Jew," Professor Boris Anisimovich Rubin, who took the trouble to invite me to his home in the hope that I would write his biography. Well, one must remember that somebody wrote a life of Napoleon's police minister, Fouché. . . .

I had a long talk with Rubin, who liked the sound of his own voice. It was on that occasion that he said: "Remember what happened to Rapoport." Yes, I do remember, and I greatly wish that people in my country knew more about him—not as the father of chemical mutagenesis, but as a man of integrity and a martyr to science.

CHAPTER 7

CITIES AND PEOPLE

"Not only was it impossible to have a human conversation with anyone, but there was also a great shortage of good beef."*

It will be recalled[1] that Professor Bursky, vice-president of the Lenin Academy of Agricultural Sciences, expressed the opinion in the early thirties that a state farm was a perfectly suitable place for training researchers and promoting the development of science. He was not saying anything original in this, but merely pushing to its logical conclusion the Soviet regime's favorite principle that collective work is always preferable to individual effort. The idea that scientific thought could be "speeded up" by concentrating the brainpower of a special group on a given problem was a simple adaptation of the experience gained, for example, in digging trenches. Far from dying out as sophistication increased, this notion has become more and more influential over the years. I have already spoken of the secret bacteriological laboratory that functioned at Suzdal from 1930 to 1937. During the Nazi-Soviet war these concentration camps for scientists, known in Soviet slang as *sharashki,* became extremely numerous. In the establishment directed by Academician Tupolev, fighter aircraft were designed and constructed by a contingent of full and corresponding members of the Academy and doctors and candidates of science—several dozen in all. Another *sharashka* has become widely known from the description in Solzhenitsyn's *The First Circle.*

The idea of stepping up scientific progress by "concentrating brainpower" can be traced throughout the activity of Stalin and his successors. It fits in admirably with the doctrine of manipulated science, complementing and extending it. It is far simpler

* M. E. Saltykov-Shchedrin, *Complete Works,* Vol. 2, p. 72.

and more convenient to tell scientists what to do when they are collected under one roof and subject to the same general supervision.

The first group of scientists to have this advantage brought home to them were the physicists. After the war all research connected with atomic fission was placed under the authority of the KGB. Lavrenti Beria, its chief, became the patron and boss of nuclear physics, and Bursky's "scientific state farms" took shape in the form of atomic research cities known by post office box numbers. Although marked on no map, these are real localities, and to this day they are scattered all over Russia from Moscow to the Volga, along the Urals, and in Siberia.

As to the principle of secrecy on which they operate, we may safely apply Norbert Wiener's dictum that "it is both far more difficult and far more important for us to ensure that we have . . . an adequate knowledge than to ensure that some possible enemy does *not* have it. The whole arrangement of a military research laboratory is along lines hostile to our own optimum use and development of information."[2] For the moment, however, we are not concerned with secrets, valuable or otherwise, but with the conditions and moral atmosphere in which hundreds of candidates and doctors of science live and work in these clandestine centers. Unfortunately it is hard to get information about them: all those who have lived in these barbed-wire enclosures, half labor camps and half arsenals, have been so effectively intimidated that they prefer not to utter a word about their experiences. However, I was able to record an account by someone who lived in a "science city" as a child, and I quote his impressions in full.

My father was an atomic scientist who lived for many years in a science city in central Russia. The settlement was surrounded by wretched villages and protected by a barbed-wire fence no less impressive, I am sure, than the one which protects the borders of the Soviet Union. The chief thing I remember is the strictness and even cruelty of the hierarchical system. The commander of the "site" and his chief specialists lived in palatial, fortresslike houses; heads of departments (like my father) had villas in a Stalinesque style, with verandas and little courtyards. Candidates of science lived in small uniform houses like cottages in Finland, engineers in apartment blocks, and service staff in barracks. I saw the inside of these for the first time after a maid of ours married a mechanic: they shared a room with another couple, the space being divided by a cloth curtain.

But there were people in the "city" who lived much worse, namely the prisoners in the labor camps which surrounded it on every side. Every day at dawn for an hour and a half I used to hear the tramp of convicts being led out to work under military escort. Occasionally, though very seldom, one or another member of this faceless mass came alive as an individual. Once a team was set to work digging a sewer alongside our home—I remember their completely shaven heads and gray, exhausted faces. My mother kept a barrel of salted cucumbers in a passageway, and the starving convicts fell on it and devoured the lot. As long as the prisoners worked in the narrow lane, soldiers kept guard at each end of it and in the middle. One day, however, coming home from school, I saw our front gate open and a shaven convict come out: he was cheerfully swinging a mess tin and made toward the end of the street. One of the guards was asleep on a pile of logs, but his companion, without a word, took aim and fired. The convict dashed behind the house, and the soldier nonchalantly resumed his rest. I waited till the shooting was over and continued on my way home unperturbed. Incidents of that sort surprised nobody during those years (1947–1953). After Stalin's death the convicts disappeared and their place was taken by men from construction battalions, but the spirit of the place was just the same. In these military-scientific-industrial complexes, class distinction and secrecy are just as rigid nowadays as they were under Stalin.

We never got any reliable information from the outside world. The scientists who made the bomb were convinced that they were working in a good cause and that, but for them, the American imperialists would already have attacked the Soviet Union. Even the convicts firmly believed that America was a predatory country where the workers lived in direst poverty. I remember, in our city there was a tumbledown shanty with a hole in the roof—an old man and his wife lived there, both wretchedly poor. A convict walked past it and remarked: "What a way to live! Might as well be in America!"

Boys at school reacted to the class system in a more direct and simple way than their elders. Fights were constantly breaking out between us children of scientists and the ones from workers' and technicians' families—these were openly class conflicts and often turned into real battles. The ragamuffin children used to shout "Death to the beavers!"—that was their parents' nickname for people with university degrees.

Such was life in a settlement of physicists; but the biologists also have centers of their own. Although the Soviet government signed the Geneva convention against bacteriological warfare in 1925, from the late twenties onward there was a secret institute near Moscow, the first director of which was Professor Velikanov; it was later moved to an island on Lake Seliger in Kalinin province. When the war broke out, Stalin had it transferred farther inland to Kirov (Vyatka) and even designated the building it was to occupy—namely the hospital in which he himself, as Soso Dzhugashvili, had once been a patient before the Revolution. In those days the hospital was on the fringe of the town, but meanwhile Kirov had expanded to such an extent that the former hospital with its concrete enclosure and its store of deadly infection stood in the very center, a stone's throw from the party's regional and executive committee offices. The general layout resembles a set of nesting dolls: the city of Kirov is on the outside, then comes the Bacteriological City guarded by regular troops, and then, within a further enclosure, the actual production unit, protected for greater safety by troops of the KGB.

The Kirov center is a highly attractive place to medical researchers, microbiologists, and biochemists. A candidate of sciences and senior research officer may earn 600 rubles a month, i.e., more than a head of department with a doctor's degree in the neighboring Kirov Medical Institute. While the shelves in Kirov shops are generally empty, the inhabitants of the Science City can obtain everything they need. True, their lives are governed by the strictest discipline and secrecy, but most of them bear this cheerfully in return for all the benefits that a senior scientist enjoys—a place to live, a six-hour working day, an annual stay in a comfortable holiday resort, the opportunity to defend a thesis on a secret subject with little trouble to himself, and so forth.

A doctor of sciences who has worked for many years in Bacteriological City says that his docile colleagues not only know that the weapon they are making is illegal but are also aware that it is ineffective. It is, in fact, a bogus weapon, at best fit to win a war in, say, Mali or Mozambique. Nonetheless, 125 researchers continue to work at the center, deceiving their scientific and military bosses for the sake of personal convenience and privilege. "The atmosphere is one of envy, fear, and suspicion," my informant continued. "That is the tradition here. Velikanov, the first director, and his wife were shot. Kopylov, his successor, was found dead in his apartment—no doubt he committed suicide to escape arrest, or else they just finished him off there. Suicides are quite common, but the secrecy is such that one never knows the reasons. Nevertheless, a majority of inhabitants enjoy the life. They are all afraid of losing the benefits that belong to their work, and so they do their best to undermine the position of potential rivals. It's a fight of all against all, and the chief weapon is again dictated by the system of secrecy—you have to prove that your rival is lacking in security-consciousness. That's the best way to damage him, and so the whole population plays at what they call 'vigilance.' This produces all kinds of situations. Colonel S., a doctor of sciences, for instance, dropped a permit from his pocket; Colonel P., a candidate of sciences, was walking behind him and picked it up. Instead of returning it he sneaked off to a permit office, so that S. got into trouble. In another case some enemies of research officer R. stole two pages of a thesis he had just written on a secret subject, knowing that it would land him before a court martial."

What do the scientists of Bacteriological City think of the moral aspects of their work? As a rule, they don't discuss them among themselves, but my informant had heard some views expressed in intimate conversations. Most of his colleagues quoted Lenin to the effect that if your opponent possesses a certain weapon it is the height of folly and imprudence not to equip yourself with it. This is a favorite saying of Colonel General E. I. Smirnov, chief of the Seventh Department of the General Staff of the Soviet Army, who is a doctor of medical sciences and commands all bacteriological centers. Those who share his simple view point out that the Americans have their Fort Detrick where similar weapons are prepared,* so why should the Soviet government hesitate?

Others justify themselves on the ground that they are military men under orders and do whatever work they are assigned to. Once bacteriological weapons are produced, it is the government which decides whether to use them or not—so they, the scientists, bear no responsibility.

"For a long time," my friend continued, "I accepted the argument about being under orders. It isn't easy to leave the army, and

* According to Soviet propaganda, Fort Detrick allegedly is the United States center for the manufacture of biological weapons.

even less so to resign from a secret establishment. But once we had what you might call a 'pure experiment.' A big department was closed down in 1965. The rule in such cases is that the personnel concerned may apply for a different job or even to be demobilized—but, out of some dozens, exactly two took advantage of it; the rest had no intention of leaving the gold mine. So they have no moral scruples about going on breeding lethal cultures and billions of insects carrying infectious diseases. And Kirov isn't the only place of its kind—there are plenty more."

But let us leave aside the atom and hydrogen bombs and bacteriological centers with their inhuman and lunatic purposes, and look at scientific cities of a more ordinary and more or less open type. These began to be constructed in the late fifties, one of the first being the well-known Akademgorodok, near Novosibirsk. The idea originated when Khrushchev, the first head of a Soviet government to visit a foreign country in peacetime, was introduced to the wonders of American university life. He was greatly taken with the green lawns and elegant campus buildings, and his advisers impressed on him that this was where major scientific work was carried on, to the great benefit of the country. Khrushchev returned home fired with the ambition to "catch up with and surpass" America, not least by constructing scientific cities.

The Russians, however, modified the American formula in important respects. In the first place they gave the scheme a political slant, without which no undertaking is thinkable in the Soviet Union. The scientific centers in Siberia and, later, in the Far Eastern Region were conceived as outposts of colonial rule. M. A. Lavrentyev, the first chief of Akademgorodok, said more than once in private that it was necessary to take precautions to ward off the Chinese threat, not only in the military but also in the social sphere, by implanting Soviet civilization more deeply in the hearts of the 30 million inhabitants of Siberia. This, according to him, was the main purpose of the Novosibirsk center.

The center also provided valuable ammunition to the Soviet propaganda machine. In a country where bears roamed until recently and which the government (the Tsarist government, of course) used as a dumping ground for revolutionaries, the Soviet regime had created a science city—yet another proof of the benefits of socialism!

The Moscow intelligentsia also had its idea of the advantages that might ensue. I remember an acquaintance, a forerunner of the dissidents of later years, arguing fervently that "whatever the government's own purpose may be, the concentration of thousands of intellectuals in a single small town is bound to have a tremendous effect. In a forcing house of that kind, a whole new philosophy of Soviet life may suddenly develop."

The original plan was to build the centers away from large towns, but here the facts of Russian poverty asserted themselves. Owing to the chronic shortage of foodstuffs, the centers tended to attach themselves to old-established industrial cities where at least some kind of provisions were available. The next problem was how to attract skilled and active scientists from Moscow, Leningrad, and Kiev. It would have been simplest to draft them forcibly in the time-honored way, but the new "liberal" epoch called for a different plan. It was therefore decided to lure the best brains by providing living space, the most sought-after benefit of Soviet life. Villas and apartment houses were built, and further inducements were offered in the form of rapid promotion, facilities for defending doctoral theses, and the promise of early membership in the Academy. Similar methods were used to attract younger people to scientific centers in the Moscow area such as Obninsk, Dubna, Chernogolovka, Pushchino, Protvino, Zhukovsky, and Zelenograd.

What happened to the scientists who were thus enticed by the promise of professional and material advantages? People who lived at "Novosibirsk-Scientific" in the fifties and even in the early sixties say that the atmosphere in institutes and laboratories there was indeed more congenial than in Moscow. In the streets and squares and on skiing trips you really found scientists engaging in lively discussion of their work, of developments in art, and so on. It also appears that during the first decade, that of the post-Stalin "thaw," relations between senior and junior staff were comparatively democratic. All of them, from doctors to simple graduates, had to buckle down to the job of constructing their own laboratories as quickly as possible, and this united the generations. There was also a genuine blossoming of cultural life. Poetry, music, and film clubs sprang up, and a café known as the "Integral" served as a meeting place where scientists could make the acquaintance of the country's foremost writers of prose and verse, as well as *chansonniers* who sang their own compositions to the guitar. Another celebrated literary center was a bookshop with

the romantic name of "Granada." Apart from this, the staff of
the newly founded institutes produced many interesting scientific
ideas and discoveries. Physicists and mathematicians were much to
the fore, but so also were geologists and specialists in hydrody-
namics.

In those days the scientific center was much written and talked
about. The "creative intelligentsia" of Moscow took many opportunities of visiting it. Yuli Krelin, a doctor of medicine and a
fellow author of mine who was at Novosibirsk in 1968, wrote a
description which well conveys the sense of fascinated reverence
that the center inspired in many of us:

> As you drive around the city you marvel and rejoice at the beauty
> of everything—the houses among the conifers, the central complex
> consisting of the hotel, post office, shops, and film theater. All round
> the outskirts are delightful-looking villas inhabited by academicians
> and doctors of sciences. And the institutes, the nerve centers of the
> place, are beautiful too. It is hard to say what is so attractive: the
> buildings are the same as elsewhere in the country, the homes,
> shops, and places of work are no different, but everything is beauti-
> ful nevertheless. It may be because of the woods, or because you
> feel that all these common, everyday houses are blessed with spirit
> of their own—a sense of truth and of the future, a spirit of science
> and of the intellect. . . . You walk about the city, into the shops
> and among the crowds—yet "crowd" is the wrong word. . . . I
> have never seen such an almost unbroken succession of intelligent,
> sensitive faces. The feeling grew on me that every woman I saw was
> beautiful and that all the men were clever, athletic, and handsome.[3]

When Krelin visited Akademgorodok, it was already past its
peak; by the late sixties a general change had set in. The forests
and villas were still there, squirrels still hopped across the streets,
but the spirit had gone out of the place and it was rapidly losing
its original atmosphere. The decline was something like that which
in the past afflicted utopias such as Fourier's "phalansteries" or
the dreams of Vera Pavlovna in Chernyshevsky's novel *What Is to
Be Done?*

The first trouble with the scientific cities was, as I have already
mentioned, a shortage of foodstuffs. This in turn immediately
highlighted the class divisions that prevailed. Junior scientists were
allocated one type of food coupons, doctors of science another,

while academicians received "Kremlin" rations. The most serious shortage was of meat. As a junior researcher walked home from the butcher's with the modest piece of beef to which his ration entitled him, he would see a van drive up to an academician's villa and a pair of strong young men lift out heavy hampers covered with napkins under which were concealed choice meats and other delicacies. There was also a special academicians' club with a first-class restaurant, and it appears that in the club reserved for ordinary scientists there was a serious discussion as to whether "candidates" might be allowed to use it as well as doctors.

The highly stratified system that thus developed led to a sharp decrease in mutual trust. Scientists who already tended to associate only with others in the same discipline now began to stick to their own "class" as well. "Everything is cut and dried," said sociologist A.A. on the first day of my stay at Akademgorodok. "Candidates have their place and doctors have theirs. As for academicians, they are invisible and inaccessible, like gods." Juniors no longer got a chance to discuss scientific problems with their elders on equal terms, let alone social problems. Hence there was a wave of unbridled careerism, with everyone doing his best to scramble up on to a higher rung of the ladder as fast as possible.

The unprincipled character of social relations spread to scientific matters as well. The atmosphere was well described by academician and physicist Lev Andreyevich Artsimovich (1909–1973): "A peaceful, quiet life in conditions of extreme specialization and complete indifference to what one's neighbor is doing—this, unfortunately, is common form in a good many of our institutes. It is about as hard to achieve important scientific discoveries in such an atmosphere as to buy Aladdin's lamp or a magic wand in a Moscow department store."[4] And indeed the performance of the Akademgorodok institutes has been falling off for the last decade, even as the number of scientific "cadres" increases year by year.

Within ten years of the foundation of Akademgorodok, the quality of its staff took a sharp turn for the worse. Researchers who had made a name for themselves and achieved higher degrees gravitated to Moscow, partly owing to the changed social atmosphere and partly because of the shortage of consumer goods, including the same old problem of beef. As a former inmate put it to me, half in jest: "I slog away at work, come home to lunch, and my wife says: 'Sorry, Yasha, there's no meat, have some bread pudding.' And it's like that for a year or two, maybe seven years—how long do you think one can stand it?"

There was another reason why the quality of staff fell off. When Akademgorodok sprang up alongside Novosibirsk, hundreds of people in the town—engineers, chemists, teachers, and doctors from local factories, schools, and hospitals—flocked to the scientific institutes in the hope of one day securing a doctor's degree and a higher standard of life. Naturally they differed widely in their talents and qualifications. Local sociologists managed to study the process, and even published a few figures when the political climate became milder. The conclusion they came to after conducting a poll was that 67.9 percent of recruits to the institutes had enrolled for genuinely scientific reasons but that 32.1 percent —nearly a third—had done so in the hope of material advantage.[5]

The crucial year for Akademgorodok was 1967, when forty-six researchers from various institutes signed a letter of protest to the Central Committee of the Communist Party of the Soviet Union against the arrest and imprisonment of the poet Yuri Galanskov.[6] Most of the protesters were young, and nearly half were Jews. The authorities' response was to stage mass repressions and inspire acts of anti-Semitism. Public acts of penance were demanded of the offenders, and, for those who refused, conditions were made so unpleasant that they had to leave their jobs (biologist R. Berg, mathematician A. Fet, writer I. Goldberg, biophysicist Zaslavsky, and others): those who had already composed doctoral theses were not allowed to present them. The Integral café and the poetry and music clubs were declared hotbeds of disloyalty and were closed down; meetings with writers and singers were stopped, as were exhibitions of unofficial art.

The government took special pains to incite senior scientists against junior ones. Academicians S. L. Sobolev and A. D. Aleksandrov, both mathematicians, signed a letter to the Central Committee urging stronger measures against dissident youth, while geologist A. A. Trofimuk, another academician, proclaimed: "Our city is a beacon which a few young snivelers have been trying to extinguish." Two years later, in March 1970, I heard Academician G. I. Budker, director of the Institute of Nuclear Physics, say: "We built Akademgorodok as a home for science; these youngsters are interfering with us and holding up the development of science, and they should be swept out with the garbage."

In March 1969 I went to Novosibirsk-Scientific for the first time to lecture at the invitation of the Scientists' Club. My first talk was about Academician Vavilov and his death in prison. At that time there was intense interest in the great biologist's career and the persecution he had been subjected to. Audiences of two or three hundred had come to hear me lecture about him in Moscow, Leningrad, and Saratov, but, to my surprise, not more than fifty turned up in Novosibirsk; moreover, the chairman, Corresponding Academician D. K. Belyaev, allowed no questions and closed the meeting immediately after I had finished speaking. When I asked why the audience had been so small, he replied: "We don't care to have our young people listen to this sort of thing." In other words, the juniors had simply not been allowed to hear what I had to say.

Next day there was another example of the "generation gap," when I spoke on "Why a Scientist Needs a Conscience." The seniors again endeavored to protect the juniors from being exposed to such a sensitive theme; they used the handy method of confining the meeting to a very small room, and only academicians and doctors were invited. The juniors, however, got wind of the plan and filled the room well before time, so that when the bigwigs appeared they had to put up a fight for their "rightful" places. After the lecture I heard once again from Professor Gaisky, chairman of the club, the argument that young people should at all costs be protected from hearing accounts of the tragic fate of some scientists and the unprincipled behavior of others. "They might misunderstand and draw wrong conclusions," said the professor reprovingly. This man's father, I may recall, was an eminent microbiologist who was lucky to escape death when working in a scientific *sharashka* in the 1930s.[7]

Akademgorodok went through a long period of antipathy toward young people, and apparently it still suffers from this senile affliction. Here is a typical dialogue between a doctor of sciences and a visiting writer:

"Somehow I don't see any outstanding talent among our young folk."

"But how much do you see of them?"

"I lecture to them."

"That means you have no direct contact."

"I haven't time for anything more—I can't give courses and seminars, I'm too busy."

"Then why don't you just take a single group?—listen to their voices and find out what they're thinking about. Maybe you'll find they aren't so stupid after all."

"Why are you so keen on defending the young? You seem to forget that everything bad has been the work of young people."

"In the first place they have done good things too, and secondly they've done everything that's new. But if they *are* responsible for bad things, it's all the more reason to see more of them and teach them ideas of goodness and integrity. It's when they come to value gain more than moral standards that they start behaving badly."

"But that would mean working every second of every day, and how am I to find the time? It's impossible!"

"All the same, it may be the most important thing you could possibly do."

However, the opposition within Akademgorodok was not exclusively between the old and the young. In the late sixties and early seventies one heard expressions of social unease from scientists of the older generation, not to speak of arguments between young physicists and young biologists which ran, for example, as follows:

"All right, you physicists have given much to the world, but you have plenty on your conscience too—all those destructive weapons. . . ."

"We are scientists first and foremost, and we can't shut our eyes to the truth when we see it. The bomb is just a by-product—you always have to pay a price for discovering the truth."

"I agree, you can't hush up something when it's been discovered, but doesn't your conscience bother you? If the world is to be saved it can only be through conscience and morality."

"What's all the fuss about? The bomb? But if it wasn't for that, governments today would have to keep millions of people under arms. The world would be nothing but an armed camp. We physicists may be the ones who've saved the world. Not without expense, of course. . . . But you geneticists would do far worse things if you had the chance."

So the physicists have a clear conscience, and it's the geneticists who should worry.

Here is another typical dialogue from Akademgorodok in the early seventies:

"Do you see the scarf that waitress is wearing?"

"How on earth did she get it?"

"How indeed? They were only on sale in the closed-order shop for academicians."

"Ah, to hell with it. . . ."

"Yes, but you wouldn't believe what a fuss there was. The academicians' wives were up in arms—it was an outrage, against all the rules, those scarves were for the higher echelons."

At the beginning of 1976 I had an opportunity to hear more about life in Akademgorodok. An old friend, Dr. Sh., a biologist, arrived in Moscow, followed by two other scientists who had recently left the Science City. What was the place like on the threshold of its twentieth anniversary? The biologist:

> I spent eight years in Akademgorodok. I made a lot of friends there, but now I've left it for good. For me, as for many others, the place has lost its attraction. There's no longer a spirit of scientific adventure, it's just a factory turning out dissertations. Good, bad, or indifferent, the Academic Councils rubber-stamp them all—they can't knock their own people. What has ruined the place is the communal-kitchen atmosphere. Everyone knows everyone else, they all see one another morning, noon, and night, and everyone lives in constant fear of what the others may do to him. The monotony of visual and mental impressions has ended up by making them indifferent to science as well. Nowadays you never see people on the street, in shops, or in the club who forget everything and get absorbed in scientific discussions with one another. All they want to do after a day's work is to dive back into the burrows where they live.

What does a candidate or doctor of sciences do in his spare time? Sh. described to me what he considered a typical case: a friend of his aged forty, a doctor of physics and head of a laboratory, who would lie for hours on a sofa watching television or else go on solitary, aimless skiing trips. Social life has changed radically since the café and clubs were closed down; it is chiefly fueled by vodka and cognac, the inevitable accompaniments of every gathering. In the home, people no longer converse or exchange views; they seem to have lost the faculty of talking coherently or listening. The solitary drinker has arrived and taken over. Local sociologists estimate that no less than 35 percent of men and women in Akademgorodok get drunk habitually, many of them daily. Another of Sh.'s friends, a doctor of geology, maintains that true friendship and love have become an impossibility. He points to the innumerable adulteries resulting from boredom which have become a local way of life, with men and women openly boasting of their bedroom escapades.

Sh. believes that all the most creative personalities have already left Akademgorodok, while the institutes and the central administration are run by smart operators, ruthless and crafty "politicians" to whom science is simply a way of doing well for themselves.

Sociologist K., who has also observed conditions at Akademgorodok over many years, adds the following reflections.

In our country the game of high science competes with morality. The vast majority of researchers hold that the pursuit of science justifies any immorality on the part of an individual. How did this attitude come about? For years our propaganda kept telling us that Soviet science would, sooner or later, solve all the problems of life. Again and again we were told that the scientist working on some grandiose scheme is a hero. To this day books and films keep on depicting the sort of character who goes without sleep and refuses to take a holiday so that he can complete some project or carry out a vital government commission on time. Certainly there are people among us who toil away like that. But even the least intelligent of them can see that as soon as science lays a golden egg, the government swoops down and carries it off. Scientists are well aware that their most exalted plans are used for the pettiest political ends. But if they were to carry this thought to its logical conclusion they would have to give up science or become politicians, and the great majority of them are incapable of either. The third choice, which suits nearly everyone, is to proclaim that science is ethical per se and that anyone engaged in science is of necessity a highly moral being. The scientist creates a useful product, and his behavior is thus indubitably moral. Many scientists today are led willy-nilly to this convenient conclusion. Thus science is victorious over morals.[8] Akademgorodok is a place where this point of view has come to prevail absolutely. Unfortunately, however, it is too late to serve any purpose, as there is no serious science left there.

My third informant is a man without a higher degree, a former university lecturer at Novosibirsk. He is worried about the generation of school pupils and undergraduates who, in a few years, will be junior and senior research officers in the laboratories.

In the heyday of Akademgorodok its then chief, Academician M. A. Lavrentyev, had the bright idea of creating a special school for highly gifted children, who would thus be turned into geniuses. By a system of competitions held all over Siberia, six to eight hundred mathematically gifted boys and girls were selected and brought to the Science City, where they were placed in a private, barrackslike boarding school. Taught by academicians and corresponding members of the Academy, the young prodigies learned things that even undergraduate students of physics and mathematics have difficulty in mastering. True, on the arts side their syllabus had to be slashed in half, but what use are history and literature to a mathematician? As science was moral by definition and there could be no nobler activity than to pursue science in the Soviet Union, it seemed to Lavrentyev that the children had all the ethical training they needed. Moral tutors were indeed assigned to them, but the budding geniuses imbibed moral lessons from a different source. The rule of the establishment was that if they received a single mediocre grade on any subject they were at once expelled. The fear of being thus shamefully thrust into outer darkness was ever-present, and the result was to eliminate all the old-fashioned virtues of friendship, mutual help, and the duty of the strong to assist the weak. Instead, the dominant philosophy was that of "winner takes all"—for of course those who win are never judged. Fifteen-year-old careerists and cold-blooded self-seekers began to thrive, while their classmates with weaker nerves or kinder hearts were pushed into oblivion.

However, an even graver moral test awaited those who passed into the university. In their first year or two they had little or nothing to do, as they were ahead of the other students in any case; so they idled away their time and behaved rudely to the instructors. They disliked and despised their fellow undergraduates, who in turn kept their distance from the masterminds. Various fates awaited these educationally unbalanced, ill-brought-up, and spiritually empty young people, but to say that they enriched science or improved the social atmosphere of Akademgorodok would be a great exaggeration.

As years passed, other pernicious aspects of privileged education became apparent. The academicians lost interest in the school, and teachers were taken on who were not much more advanced than those in normal schools. Standards fell, but the graduates were as arrogant as ever. Meanwhile, by picking several hundred of the most gifted physicists and mathematicians out of the common stock every year, Akademgorodok was stripping ordinary schools of their best pupils and producing dullness and mediocrity over Siberia as a whole. The experiment launched many years ago has virtually proved a failure; everyone knows this, but they do not dare close down the "school for geniuses" without instructions from on high.

I have dwelt in some detail on life in the Novosibirsk Academic City because its ailments are typical of those which afflict other centers of this kind. They are due not only to the inveterate sickness of Soviet science in general but also to the rootlessness of such settlements, a feature they share with military cantonments or seasonal workers' colonies. Most inhabitants of the Science Cities have no relatives living near them, no elders in the family, no former teachers and schoolmates; none of them has a neighbor who has known him since he was a boy, or a first girl friend whose memory helps him to be a better and more upright character. The only restraint on antisocial behavior is native caution or the fear of losing one's job: scarcely ever the stirring of conscience due to censure by one's nearest and dearest.

Rootless cities have a peculiar atmosphere of their own—on the one hand rather drab, but on the other pervaded by an electric restlessness. I have visited Protvino, Pushchino, Dubna, Ramon, Obninsk, and Chernogolovka. They differ in their topography and architecture and in the type of scientific problems they are concerned with, but I felt that they all had something in common. Everyone feels himself to be a temporary resident, and this impermanent, unsettled existence leads eventually to inner instability, even among the ablest and, one would think, most successful researchers. We shall hear presently what they themselves say about it.

Chernogolovka, some thirty miles from Moscow, is a settlement founded in 1958 comprising three academic institutes. It is an hour and a half bus ride to the terminus of the Moscow subway. The present population is about 10,000, and it is expected to grow to 30,000. There are about a thousand scientists and a large number of young children. Apart from living quarters and institute buildings there is a single shop, a post office, a school, and the Scientists' Club. The fringes of the town merge into forest. There is no local self-government of any kind: the place is run by the academic administrator, F. I. Dubovitsky, who is also deputy director of an institute and is locally known as "Tsar Fedor Ivanovich" (i.e., the son of Ivan the Terrible). Dubovitsky, an explosives expert, is a man of stern temper and firm principles. His conception of the role of a scientist is much like that of Russia's erstwhile rulers toward the corps of musketeers: the serviceman's whole duty is to know his harquebus and powder flask, and all the rest is nonsense. Accordingly, in the eighteen years of his rule Tsar Fedor Ivanovich has not built a single swimming pool, tennis court, or restaurant, all of which he considers superfluous luxuries.

I once caught a glimpse of this autocratic personage when my hosts were showing me around the town. We were in the general store when a robust elderly man came in: he seemed to be getting on to seventy, and looked like a retired colonel. His expression and gait showed him to be the boss. A respectful murmur arose from the men and women standing in line. All commercial operations ceased at once; even the cashier stopped taking money. A white-coated manager sprang out of the interior of the store and, all but taking the honored visitor by the arm, led him into the inner sanctuary. One by one the counter assistants were summoned there, while the uncomplaining customers waited in silence.

Another sidelight on the personality of the "Tsar" came in a casual conversation in a laboratory. Apropos of a certain Dr. Kh., whom I did not know, someone uttered the significant words: "Of course Dubovitsky likes Kh., because Kh. is afraid of him."

Well, that was Chernogolovka in 1975. What did the resident scientists think of the place and of their own lives?

Aleksandr Evgenyevich Shilov, forty-five, a professor and doctor of sciences, is the second most senior person in the settlement. He lives in a two-story villa with his wife and son and a dog; he owns a car, has a direct telephone link with Moscow, and receives special supplies from the capital. In Shilov's words:

The original idea of creating a place where scientists could live in the countryside, free from city noise and cares, was clearly sensible and fruitful. For a long time Academician N. N. Semenov was looking for someone to act as his deputy here. In 1962 I came back from an official trip to England; I liked the idea of living in a house of my own, Oxford-style, so I agreed to come here.

I know very little of the local people and their private lives; my wife and I try not to mix with them at all. But we have noticed one or two changes over the past few years. There is something bleak about the place. Architecturally it's a complete failure. Those who planned and built it gave no thought to any conveniences of life for the present generation, let alone the future. By now the children of researchers who came here in the early sixties are growing up, and what are they to do after graduating from school? The only jobs here are in the scientific institutes; so, regardless of their children's tastes and inclinations, parents are sending them to chemistry and physics faculties simply and solely to have them trained, bring them back to Chernogolovka where there's a roof over their heads, and squeeze them by fair means or foul into the Institute of Physical Chemistry. The result is that the character of the staff is gradually changing. Those who first came wanted to work at science, but many of the newer ones are simply interested in getting an apartment. This is a disaster for the Institute and for science, but so far we don't have a cure for it.

Alla Konstantinovna Shilova, in her late thirties, candidate of sciences and wife of A. E. Shilov:

The atmosphere has definitely deteriorated since they built an instrument-making factory here and three thousand workers came flooding in. It used to be a clean place—no one would dare drop cigarette butts on the sidewalk, but now it's filthy. The factory is attracting more and more youngsters from Noginsk, and they jump the queue for apartments. Long-haired guys with transistor radios, and they're mostly hooligans.

G.L., forty-one, doctor of sciences; author of internationally known research papers. Married, with one child. Lives in a multi-story brick building:

My wife and I live a secluded life. I don't feel like inviting my colleagues home. Conversation is never about anything but private grievances, or women chattering about clothes and men about laboratory problems. It isn't safe to tell anyone what's really on your mind.

Outside the laboratory it's as dull as ditch water. People are constantly leaving their wives and families for the sake of a change, but the new marriages are just as pointless as the old ones.

R.G., about forty, candidate of sciences. Married, with two children. Regarded by his superiors as a keen, hardworking researcher:

I like it here. I love the countryside; in the mornings I go jogging in the woods, and in winter there's skiing. I have no time to get bored: I come to the lab at nine in the morning and I don't leave till nine at night. There's plenty of work and I enjoy it. But a lot of people do get depressed here—it's almost impossible to get out of the place, and they find that dispiriting. The Institute considers your apartment its own property; you can only exchange it for one in another city through the law courts, and that means ructions and a black mark in your personal record. It's a risky decision to take. Also, the directors will only hire young people in your place—they regard old men as dead wood.

Lab assistants and junior research officers try to get candidates' degrees simply because they're badly off. Otherwise you can be an assistant for twenty years and never make more than a hundred rubles a month. Consequently people push and shove their way into science whether they have any aptitude or not, and some of them are the kind that science ought to be well and truly protected from.

R.G. also told me that, ten years earlier, research staff used to visit one another socially, but these ties were disintegrating, as it was boring to talk shop over the teacups. On the other hand R.G. would have very much liked to correspond on scientific topics with such experts as Charles McKenna of Florida, Leonard Martinson of Texas, or the nitrogen-fixing people at Ohio State University. Charlie McKenna, especially, was a great guy with a good brain—how was he getting on, and what were his latest ideas? But R.G. is only allowed to write to the United States through the Foreign Department of the Academy of Sciences; his letters take months to get there and the replies come half a year later, by which time he has lost interest and the problem is out of date. And if you touch on anything personal in your letter, you can be quite sure it will vanish into limbo. McKenna, being a foreigner, of course cannot come to Chernogolovka, and there is precious little hope that

R.G. will ever be allowed to go to the United States. Every year he puts in an application to the Academy, but . . .

A.K., twenty-eight, recently married, a newly fledged candidate of sciences. Salary 175 rubles a month; first-class personal record. Lives with his wife in a hostel, but hopes to obtain an apartment:

> My wife is a medical doctor, and she's very miserable here. I know others too who dream of getting out: they say they're dying of boredom and loneliness. But I don't intend to leave—I like the work, and I'm used to this sort of life. My father's an army officer, and we used to spend our lives moving from one cantonment to another. In those places too no one had any local friends or relatives, so I don't mind the conditions here. I make friends with the people I work with. We get together at home and talk shop there—what else is there to talk about? There's no time to read literature, only just to look through *Literaturnaya Gazeta*. Of course it would be nice sometimes to have a break, but there's nowhere to go except the Club. That's reserved for candidates and doctors, and each one has his seat reserved in the auditorium—I haven't got mine yet, as I only just took my degree, so I rely on the odd ticket left over at the box office. . . .

Candidate of sciences E.F., thirty-two, regarded as a brilliant man with extremely interesting ideas:

> It's like living in a glass house. If I want to be alone for a while, just to think about scientific matters, there's no place I can go. Our flat is tiny, and my wife and the child are there. If I go out, everybody I come across is someone I know, and they drag me off for a drink or some time-wasting conversation. The only escape is to run off to Moscow and lose yourself in the noise and crowds. Being in a crowd is the best kind of solitude—it recharges you for a whole week. My wife finds it even harder to bear the life here, but where could we go? We've got our apartment.
>
> Public opinion here is more censorious than in a big city. If people want to indulge in debauchery and hard drinking, they do it in secret. It's considered entirely proper to stay in your lab until midnight, but wrong to take your child for a walk or just stroll about in the woods. If a restaurant ever opened here it would soon go out of business—researchers would be afraid to set foot in it for fear of offending "public opinion," which really means their superiors.

Mrs. V.L., the wife of a doctor of sciences and herself a physician, has problems no less serious: her son, for instance, has several times been beaten up at school and has had boys shouting "Yid" at him.

The school at Chernogolovka is a somewhat peculiar institution. On the one hand it is used by the Academy of Pedagogical Sciences as an experimental center: experts from Moscow try out new teaching methods here, and the architects who designed the school even received an All-Union prize. At the same time, it is a kind of container in which two strains of pupil have coexisted for years without intermingling or merging: the scientists' children and the children of peasants from nearby collective farms. These two groups are at daggers drawn: the children of the intelligentsia, who are brighter and have read more, assert themselves in class, while the others get their own back in recess and after school. Fights, quarrels, insults, and name-calling are the staple of school life, and there is nothing any Academy of Pedagogical Sciences can do about it.

Another of V.L.'s worries is what to cook for dinner each day. The local store is poorly supplied from resources in the immediate area. The forty top scientists get their groceries "on order" from Moscow. As V.L. has neither a car nor the right to a special ration, all she can do is to help her husband on with his rucksack and send him off by bus to forage in the Moscow shops. As for living in the countryside, yes, she likes it, but somehow conditions are such that she hasn't the strength to enjoy it.

Such is Chernogolovka at the present time. As may be seen, the atmosphere and way of life do not differ greatly from those at Novosibirsk. Chernogolovka is an Akademgorodok that never experienced a "golden age": there was no brief period of social and scientific animation of the kind that left people in Novosibirsk with pleasant memories and illusory hopes. As for Obninsk, Protvino, Pushchino, and Ramon, no one there has any hopes at all. Internal relations, stripped of all pretense or embellishment, are openly based on inequality and the denial of freedom. A scientist's prestige is generally determined not by his ability or creative potential, but by his salary and living quarters, whether or not he has the right to an armchair in the Scientists' Club, and, as often as not, simply by the size of his meat ration.

Ethical attitudes are also much the same in Siberia and in the

Moscow region. I was able to form a good idea of these during the ten years I spent lecturing in various scientific cities about the moral problems of Soviet scientists, with particular reference to Vavilov's sufferings under the Stalin regime. Prior to this I had given similar talks in the universities and research institutes of Leningrad, Tartu, Saratov, and Vladivostock; but the result was that I was soon denounced, on grounds of political disloyalty, to the Writers' Union and the Procurator-General's office of the USSR. I was obliged to give an account of myself to General Ilyin, the official supervisor of Moscow writers, and General Terekhov of the procurator's office, and was banned from addressing the Writers' Club and the Academy of Agricultural Sciences. However, at that time interest in Vavilov was on the increase, and the Scientists' Clubs at Novosibirsk, Dubna, and similar centers started to invite me direct, bypassing the Writers' Union. It was in this way that I became friends with numerous inhabitants of scientific cities.

To be frank, I found it hard to understand at first how it was that my kindly hosts still dared to invite me after I had been banned from Moscow libraries and research institutes. However, all was explained on my first visit to Dubna. When I warned the head of the Scientists' Club—Oleg Grachev, a former(?) major in the KGB—that I would be touching on some delicate points in my account of Vavilov's career, he smiled indulgently and replied: "In this place you can talk about anything you like. Camps? Prisons? How Lysenko did Vavilov in? Go right ahead, we're not afraid. Everyone here gets a fat salary."

Far from everyone in Dubna actually does receive a high salary, but the hope of doing so seems to act as a strong damper on political passions. Nonetheless, Dubna is a stimulating enough place to the outsider on his first visit. You are given a room in an excellent hotel, they show you the synchrophasotron and tell you all about rays and megavolts. You take a walk along the low bank of the Volga and lunch at quite a good local restaurant; in the evening you give your talk and then enjoy a late supper with your new friends. Seen thus, the place may indeed appear an intellectual haven; and the same is true of one's first-day impressions of Protvino and Pushchino. In Protvino you are struck by the imposing Scientists' Club and the interesting design of the local market. The young physicist appointed as your guide tells you excitedly about the houses with duplex apartments that are being built for doctors and laboratory heads. "The latest thing, I can tell you! Just as good as England!"

The center at Pushchino, where the biologists live, gives the impression of being unfinished and somewhat empty, but the beauty of its natural setting delights the eye. It stands on a height overlooking the Oka River, whence there is a fine view of birch forests extending for miles; below, the river can be seen encircling the town like a sickle. The air, too, is remarkably pure. When you have breathed it to your heart's content, you can enjoy fish delicacies at a restaurant with an appropriate Greco-Roman name (Neptune, Poseidon). The remainder of your program follows the same course as before: a lecture, intelligent conversation, a first-class hotel.

Whoever leaves this abode of the gods next morning and does not return will take with him only memories of delectable hospitality and a beautiful countryside. If you come back to Pushchino a second or a third time, however, and if you make friends there, you will get a different impression of the center with its five Academy institutes. Take, for instance, the lights in the windows late at night which you were shown proudly on your first visit as proof that "the scientist is aglow, day and night, with enthusiasm for his work." After a day or two you will discover that the only burners of midnight oil are laboratory assistants and junior researchers; candidates of sciences are in bed by eleven, and doctors even earlier. Those who have not yet taken higher degrees cannot afford to sleep: they have to grind away at their theses, otherwise they will starve.

Another local boast which visitors are impressed with on their first day is that "many researchers spend whole weekends in their laboratories." After a day or two, this undoubted fact will appear in a new light. As a woman mathematician who had lived in Pushchino for several years explained to me: "Yes, many researchers do go off to their laboratories on Sundays. But why? In our little community, family affection dries up more quickly than in a large city. A young scientist gets bored sitting at home, and domestic arrangements aren't easy; so the men use science as an alibi for not helping their wives. They have a pleasanter time in the lab, they can sit and chat with their friends all Sunday instead of working. It's a sort of intellectual tavern, if you like."

Private conversations with scientists at Pushchino reveal interesting details of their personal and professional lives. R., a young candidate of physico-mathematical sciences, regards it as the freest and most convenient of the scientific centers near Moscow. Public transport links with the capital are better, and some of the bosses are very decent people. There is, for example, Academician Aleksandr Sergeyevich Spirin, director of the Protein Institute (at which R. is employed): so far he has not signed a single letter against Sakharov, and disloyal people do not usually remain on his staff for long. On the other hand, R. considers that the Institute personnel are in general a poor lot intellectually. "We have a good library," he told me, "but the physicists, biochemists, and biologists whom I know only want to read detective stories and science fiction. Whenever I go into the library I hear them asking for 'a bit of light reading.'" There is practically no demand for serious books, and the great majority of scientists take no interest in belles lettres.

A third-year biochemistry research student told me about another aspect of life in Pushchino. A certain Elkin, a research student under Academician Gleb Mikhailovich Frank of the Academy Institute of Biophysics, cribbed his thesis from an article in a foreign journal and, to prevent exposure, removed the journal from the library and destroyed it. The cheat was only discovered two years after he had successfully defended his thesis. His young colleagues were indignant and demanded a public inquiry, but Academician Frank and the other professors preferred to hush the matter up. When it was raised before the Academic Council of the Institute of Biophysics, the necessary two-thirds majority vote to strip the impostor of his degree could not be obtained: the doctors of sciences were afraid of offending the Institute director. Young scientists continued to call for an honorable decision, but their voices went unheard in Pushchino. "We shall go on fighting against the protectors of this pseudoscientist," said my postgraduate friend, adding, however, in the same breath: "but I don't expect we'll get anywhere. This Elkin is very well in with local party people, he's a delegate to the Komsomol Congress and even chairman of the Young Scientists' Association. And who are we? Nobodies."

Protvino, a small town twelve miles from Serpukhov, also reveals its true nature on the second or third day of the visitor's stay there. Like Pushchino it is constructed of many layers, its nucleus being the synchrophasotron, said to be much more powerful than the one at Dubna. Around this diabolical merry-go-round is a ring of houses containing one- and two-story apartments, garages, shops, and nurseries, not to mention scientific careers and scientific squabbles. The life of Protvino's 10,000 inhabitants is determined by the operational rhythm of the High Energy Institute, and the Institute plus the town are part of a still more august organization: the State Committee on Atomic Energy under the Council of Ministers of the USSR. Geographically Protvino lies on the Protva, a tributary of the Oka; politically it stands at the point of intersection of the state's principal ambitions as a nuclear power: atomic terror and atomic blackmail.

Everything here is calculated to induce the young scientist to sell his soul. On arrival at Protvino he discovers that, however wretched life may be in the surrounding villages, supplies to the nuclear research center are always adequate. He is told that in two or three years he will be given a one-room apartment to himself, and he may in due course aspire to a duplex, provided he works hard, does what he is told, and is careful not to give away his real thoughts. Life, of course, is insufferably boring; all topics of conversation have long since been exhausted, and people stopped visiting one another long ago. A junior research officer cannot maintain his wife, let alone a child, on his salary, and there is no work for his wife except as janitor or cleaner. Nor can a junior researcher expect a doctor of sciences to strike up an informal conversation with him on the street or in the club. Nevertheless it is also true that Protvino is a place of opportunities, a springboard city. Eventually, in six or seven years' time, the young researcher will be able to quit this godforsaken hole in the woods and head for Moscow, and that is certainly what he ought to do. Meanwhile not a word or gesture out of place, because the atomic city is a place of intensified surveillance—every second person is an informer.

After I gave my lecture at Protvino (once again on "Why a Scientist Needs a Conscience"), the friendly young people asked me —not in the lecture hall but later, over a glass of beer, and in a spirit of curiosity, not of passion or irony—why exactly a scientist had to be an intellectual in my sense of the term. Not one of them agreed with the Oxford definition of "intelligentsia" as "the part of a nation that aspires to independent thinking."* Independent thinking? The youngsters snorted derisively. In their work they had to check every idea that came into their heads with the laboratory chief (who in return granted each of them a slice of the cake and allowed them to present a doctoral thesis). As for social or, God forbid, political ideas, let someone else do the thinking and talking: scientists were not allowed to, and in any case they had no time.

The young people did ask me what I thought of Academician Sakharov, and they seemed glad to know that I respected him as a scientist and a public figure. But their own feeling was that politics had nothing to do with science. Take the laser, for instance, an invention developed in both the Soviet Union and the United States: both sides pumped in roughly the same amount of money and took about the same time—what had their different social structures to do with it? An electrical engineer who talked on these lines, a pleasant young man, smiled at me condescendingly. He was convinced that science would forge ahead of its own accord. If a discovery is technologically ripe, it will be made, irrespective of how a country is governed or what the radio says. The Protvino researchers could not care less what the radio said, but they had to turn out their candidate's and doctoral theses at any cost. In this world there are rules, and by observing them you can achieve your cherished goal; the young people despised these rules at heart, but they had no intention of disobeying them.

I will end my remarks on scientific settlements in the Soviet Union by quoting a specialist opinion. A few years ago a Polish woman sociologist, Ja——ska, made a trip to the USSR, including Novosibirsk. Asked by Soviet reporters what she thought of Akademgorodok, instead of uttering the expected compliments she replied: "It is wrong to build settlements like this. We live in an imperfect world, a rigidly differentiated society. In big cities this differentiation, whether professional, economic, or legal, is blurred, so that people from the lowest social strata do not suffer their inferiority so acutely. But here in Novosibirsk differences in

---

* This is the definition in the fifth edition of the Concise Oxford Dictionary (1964). In the sixth edition (1976) the word is defined as "class of intellectuals regarded as possessing culture and political initiative; class of persons doing intellectual work. –Trans.

status hit you in the eye. A full member of the Academy lives in a villa, a corresponding member has half a villa; a senior research officer has an apartment with a three-meter ceiling height, while a junior has one with a two and a quarter meter ceiling, on a higher floor with a communal bathroom. You only have to point at the window of a scientist's home to define his social position, his rights, his resources, and his prospects."

The Polish sociologist had put her finger on the main drawback of Akademgorodok: its destructive effect on the individual. But have the bosses of Soviet science realized that their experiment with scientific "reservations," unlike those in the West, has failed? I talked about this with several high officials of such centers and members of the Presidium of the Academy of Sciences. No, they did not see the situation as catastrophic. Of course there were difficulties, but, taking things as a whole, they believe, like Professor A. E. Shilov at Chernogolovka, that the troubles of Soviet Science Cities are simply growing pains. Their own two-story villas with garage and hot water go a long way to justify the whole scheme. As for the party manipulators of science, the system suits them down to the ground. If you ask them why it represents all that is best in Soviet science, they will answer with vague phrases about cross-fertilization and the atmosphere of enthusiasm that prevails at Novosibirsk, Dubna, or Pushchino. As we have seen, the enthusiasm has vanished, but there is something else that the bosses really do value. The researchers in Science Cities are even more dependent on the authorities than their colleagues in Moscow, Leningrad, or Kiev. Any display of private or public initiative is still less possible in these communities than in the major cities; public opinion has been reduced to nil, and the personal element in science is completely eliminated. In other words, Soviet science in a scientific city is more manipulated than anywhere else in the country—and this is what matters to the government more than anything.

CHAPTER 8

THE ETERNAL CHOICE

"They sat there quietly, and I had the impression they were whispering the usual brief prayer of educated people: 'Lord, let this pass from us!' "*

Academician Vavilov was in the habit, when he met people who interested him, of asking them, "What is your philosophy?" He meant primarily their views on science, but he was a sociable man and was delighted if they also told him their way of thinking about life, learning, and social problems. This traditional passion of the Russian intelligentsia for discussing political, religious, and philosophical issues, even with people they did not know very well, naturally fell into abeyance in Stalin's time. In the mid-fifties, after the dictator's death, it was still difficult to induce people to talk openly about their philosophy of life, and the situation has not changed very much today. One can hardly expect frank confessions in an era when telephones are tapped, private correspondence is intercepted and read, and intellectuals are recruited as informers. Nonetheless, I devoted years and even decades to trying to find out what the average member of the scientific million believed or did not believe, what was his moral philosophy, and what were the ideals of a prisoner of manipulated science.

Naturally one cannot look inside a million hearts, but during my years as a writer I came up against hundreds of scientists in all kinds of jobs, men and women whose characters and talents differed as widely as could be imagined. I met many of these people during my travels about the country, and corresponded for years with some of the most interesting ones. Some out-of-town scientists became friends of our family, and visited us in Moscow when they turned up from Krasnodar, Vladivostok, Leningrad,

* M. E. Saltykov-Shchedrin, *Complete Works,* Vol. 7, p. 179.

Tallinn, or Tashkent. I look back now with deep gratitude to those meetings on Academician Pavlov Street, free from the bustle of official journeys and long after the first shy beginnings of acquaintance, when we were able to have heart-to-heart talks in a small room lit by a single green-shaded lamp. I pay no attention to gradations of scientific rank, and so my guests always felt at ease, whether they were professors and academicians or laboratory assistants and junior research officers. All alike were happy to throw off the mental and spiritual constraint that Institute life imposes on scientists, whatever their rank and duties. For a while they became their natural selves again and, in the true old Russian style, could unburden their hearts of anything they chose.

What subjects my guests talked about during those cozy evenings! From family affairs to matters of state, from religion to academic scandal. An academician would confess with remorse that he had been made to sign an unworthy letter on some public issue; a capable young mathematician would tell me of his plans to emigrate to Israel; a scientist back from a trip abroad would tell me enviously of the equipment he had seen, and a colleague of his would seek advice on how to face the latest witch-hunt in his department. There would be quarrels and arguments too, for although my guests were all excellent people they differed widely in their religious, political, and social views—so this too was a way of learning about the "philosophy" of the scientific million.

One important fact that came to light was that people held different views as to the very meaning of the terms "morals" and "ethics." These are generally spoken of in one breath, and according to the Great Soviet Encyclopedia they are the same thing,[1] but my friends and I agreed in making a distinction between them. Ethics, as I understand it, is based on custom and tradition and represents the way in which an individual adapts to the social group. The commandments "Thou shalt not kill," "Thou shalt not steal," "Thou shalt not covet thy neighbor's wife" are ethical precepts. Ethical and unethical acts are generally perfectly clear to observers and require no further explanation. Unethical conduct is the violation of ethical principles accepted in a particular social group. In particular a scientist, according to the ethics of his profession, must be strictly honest in carrying out experiments and assessing results. He must also behave decently toward his colleagues, not demean himself by falsifying his own reports or reviewing the work of others unfairly, not misuse his vote on Academic Councils, and so forth.

Morality, on the other hand, represents a man's awareness of himself as a personality, a human being; it expresses his personal attitude in his most important contacts with the external world. Moral principles define a man's relation to God and nature, to society and the state.

Rudolf Hess, the commandant of the death camp at Auschwitz, was a good family man, a loving, caring husband and father. At work he was considered a just, correct man, and his colleagues respected him. He imposed a work ethic on his family and, as he wrote himself, considered an upbringing of this kind to be "of decisive importance for the preservation of ethical and mental health." But are we to call Hess a moral man?

Morals may be called the humanistic core of ethics; ethical behavior may be partly or wholly immoral.

I have said much in previous chapters about the ethical or unethical conduct of Soviet scientific man. It is time now to talk about his morals, and first of all his relationship to society and the state.

It is hard to agree with Soviet propaganda when it talks about the unshakable politico-moral unity of the Soviet people. The million scientists, who are quite a good microcosm of society as a whole, cover a wide spectrum from the moral point of view. For the past half century their chief principle has been that their work belongs to the state and that it is their duty to serve the state and the Soviet people. A scientist's gifts, his education and doctor's degree all contribute to enhancing this duty; for all that we possess is borrowed from the people, received as a gift from the state. Consequently it is our duty to study, invent, and discover those things which the Soviet economy requires at the present moment. This is the doctrine hammered home in the Soviet press, in biographical sketches and books about contemporary scientists.[2] I must confess that I too have sinned by inculcating this essentially servile attitude: in many of my books I have conformed to the stereotype by assessing my subject's merits in terms of the "social utility" of his discoveries.[3]

The surrender of one's own independent personality, the idea that one is an instrument of the state and nothing more, is expressed with great clarity in the memoirs of Soviet professors and academicians, which proclaim the doctrine that the whole purpose of a scholar's life is measured in terms of political and social utility. Today, in the late seventies, this attitude is wearing thin, but from the thirties to the fifties it was indeed the view of Russia's greatest scientists. I have described one of these old-fashioned types in the person of A. L. Mazlumov (1897–1972), a plant breeder and member of the Lenin Academy of Agricultural Sciences. Mazlumov, who was responsible for the cultivation of some fifty varieties of beets, told me that he used to expel from the All-Union Sugar Beet Institute anyone who tried to carve out a "piece of scientific territory" for himself. In other words, he deprived his staff of creative independence and scientific freedom for the sake of a political and national interest. Those who were willing to work for him on this basis frequently forfeited degrees and personal recognition because, as one of them put it, "they forgot their own interests for the nation's sake."

Nowadays this kind of self-denial and self-disparagement is more typical of the provinces than of the capital, and it is more usual among women than among men. Not long ago I received a letter from Maria Levitanus, a surgeon of more than thirty-five years' standing, who wrote: "I go on working because . . . I have not yet repaid my Soviet fatherland for all it has given me."[4] The "payment" is of course metaphorical, but in the writer's eyes it is no less a sacred duty on that account. The same attitude was expressed in a letter to me (quoted in Chapter 5) by a former pupil of Mazlumov's named Iraida Popova, a candidate of biological sciences and a phytopathologist at the Sugar Beet Institute. Her protest, it will be recalled, was provoked by my simply suggesting that she ought to comply with a request from a British expert to clarify certain purely scientific details.[5]

With some scientists, this type of chauvinism is a sincere and deeply felt attitude. Loyally believing that the state's wisdom is indisputable and unchangeable, they are convinced that they themselves cannot do wrong if they obey the state's commands in every detail. The field of moral choice is thus reduced to the uttermost, as whatever decision they take is sanctified from above: the authorities know best, "they know what they're doing."

This belief in the superior wisdom of those on high causes provincial scientists (in both the literal and the metaphorical sense of the adjective) to make decisions that are sometimes more than questionable. Take, for instance, questions of the relationship between man and his natural environment. Scientists, like other persons, enjoy the sight of a vernal wood, an unspoiled beach, or a crystal stream; but this is a domestic, personal feeling, and will soon go by the boards if "superior interests" are invoked against the love of nature. Although there is a Soviet law for the protection of the environment, the authorities have a hard-boiled attitude toward questions involving the purity of the air or the conservation of the soil, forests, and waterways. The immediate needs of industry, especially nuclear and defense industry, are constantly leading to the destruction of basic and irreplaceable natural resources.

In 1966 a group of leading Soviet scientists tried to prevent the establishment of a cord tire factory on Lake Baikal. They warned in an open letter that the unique local flora and fauna would be poisoned by effluents and that the country would lose a reservoir of clean drinking water, of the utmost necessity to future generations.[6] But the party's Central Committee explained that the enterprise was of military importance, and at once all protests fell silent.

The terms "military," "defense," and "secret" belong to a vocabulary of magic spells that paralyze the will of any Soviet citizen, however highly placed he may be. A "secret establishment" is something that one must not know about, let alone discuss. Only a lunatic, in Soviet society, would question the propriety of its existence. In a rocky wilderness on the Mangyshlak peninsula in the Caspian Sea there is a secret city called Shevchenko, the factories of which use desalted water from atomic distilling plants. The hundreds of scientists who live and work there—chemists, physicists, biologists—know that these installations discharge into the Caspian a concentrate containing 310 grams of salt per liter of water. As a result, for dozens of miles around the city there are no fish and no marine vegetation—the sea has simply died.

A friend of mine who visited Shevchenko tried to talk to local scientists about the irreparable harm they were doing to the environment; but they replied that they were servants of the state and in no way felt morally to blame. Slavsky, the minister concerned, had personally given scientists at the plant a five-year exemption from any penalties under the law for the protection of the environment; so for that length of time, at least, they could go on poisoning every living thing for miles around.

It is no good inveighing against the morals of ministers, but, as Yosif Brodsky put it, "politics corrupt morals, and that is our business." Some thousands of miles from Shevchenko, at Vladivostok, I myself had a chance to observe how an official can corrupt the morals of a scientist. On the morning of May 27, 1974, I was present at a joint session of the Academic Councils of Biological Institutes of the Far Eastern Scientific Center of the Academy of Sciences of the USSR. A report was delivered by the vice-director of the Institute for Biological Problems of the Northern Regions, situated at Magadan inside the Arctic Circle. I hoped the lecturer would have interesting things to say, but he proved very boring and put most of the audience to sleep. However, the ensuing discussion gave rise to an incident that I recorded in my diary.

A forestry specialist from Vladivostok named Rosenberg asked the lecturer why his Institute did nothing to improve the quality of the local soil. Magadan is in fact surrounded by thousands of miles of permafrost. In summer this melts to a depth of a few inches and the soil is covered by mosses and similar vegetation, but if a tractor runs over it a gaping wound forms in the soil and takes years to close up of its own accord. The development of local mining, especially gold mining, has turned the area into a vast marshy expanse in which nothing grows. It is of the utmost importance to rescue the soil from this condition, but the Magadan Institute had taken none of the appropriate steps. Why was this? In reply to Rosenberg's challenge the vice-director said that A. Kosygin, chairman of the Council of Ministers of the USSR, had visited Magadan in 1973 and ordered the local authorities to increase the output of gold; if they raised it by 2 percent, Moscow would provide extra funds for capital and housing construction. What was more, Kosygin had exempted them from the environment protection law—so they could destroy the soil to their heart's content so long as they supplied the extra 2 percent of gold. This being so, the Biological Institute had decided that it would be inopportune to take any measures for soil preservation.

The lecturer explained this in a completely matter-of-fact way, as though it were the most natural thing in the world; and the Vladivostock biologists stopped debating the matter as soon as they heard that the destruction of the soil had been sanctioned by higher authority. During an intermission I waylaid the lecturer on a staircase and asked him how he personally felt about it.

"It's a complicated political problem," he replied. As before, neither his face nor his voice betrayed any emotion whatever. "If Comrade Kosygin himself has decided that way, it must mean that from the point of view of state interests . . ."

"But what do *you* think about it? The soil is going to rack and ruin. . . ."

"I don't think anything. As it happens, I'm a medical doctor by profession."

It may be that the vice-director was only pretending to believe in Kosygin's superior wisdom. But what is certain is that the Biological Institute (under the direction of V. L. Kontrimavichus, a corresponding member of the Academy of Sciences of the USSR), whose profession and duty it was to protect the northern flora and fauna, did not utter a word of protest against Kosygin's illegal action. It is, after all, safest and easiest to accept the defense of "superior state interests."

I have noticed, however, that defenders of the government's wisdom are for the most part men of passive character. If they are not threatened and if there is no state official standing over them, they will not push their loyalty to undue extremes. But government control of science would be less perfect than it is if there were not room for sincere defenders of the regime, or at any rate those who wish to appear such. These include, first and foremost, administrators—the 25,000 or 30,000 party stalwarts who run the Academy of Sciences of the USSR, the Academy of Medical Sciences, the Lenin Academy of Agricultural Sciences, and the Pedagogical Academy. Other staunch defenders of the regime are the rectors of universities, directors of research institutes, heads of scientific departments and faculties. They can be relied on if only because their jobs, pay, and prestige are wholly dependent on the party authorities. They zealously pass on instructions to the million, and at times even take the initiative before they receive orders. To ensure the regular functioning of this group, no moral principles are necessary.

The second group of "loyalists" should logically consist of the pillars of orthodoxy who study and teach party history, historical and dialectical materialism, Soviet history, and literature, socialist economics, and so on. Officially, at least, the state is built on scientific foundations, and it is the function of ideologists and theoreticians to use scientific arguments to rally the population under the banner of communism. And indeed there is a large, noisy ideological army, full of spoken and written eloquence. But its books and speeches are so monotonous, dull, and divorced from reality that no one any longer pays the slightest attention to them. If there is any place where Marxism is still developing and affording real inspiration to anyone, that place is certainly not the Soviet Union. Marxism in the Soviet Union signifies nothing more or less than a livelihood for a certain number of professors and lecturers, and a course that has to be plowed through by university freshmen.

There is, however, another group of scientists who seriously contend that there is something life-giving and inspiring in the moral and ethical system of the Soviet Union. These are not professional ideologists but consist of a small number of individual mathematicians, physicists, chemists, and biologists. Thus, every now and then patriotic articles come from the pen of Academician and Nobel prize winner N. N. Semenov; the physicist Dmitri Skobeltsyn makes speeches in the same vein; the mathematician A. D. Aleksandrov and N. M. Amosov of the Medical Academy are likewise prolific publicists. All this support is greatly valued by the authorities, as it enables them to claim that top-ranking scientists are on their side. I repeat, the number of such loyal supporters who are truly eminent men is extremely small, but they are interesting inasmuch as they seek scientific arguments in defense of the regime and of its ethical system.

At the home of some Moscow friends I made the acquaintance of a member of this group, Mikhail Vladimirovich Khanin, a doctor of technical sciences and an expert on the aviation industry. I had been told that he was an honest, straightforward man, whose interests were not confined to his profession. He proved to be a man of fifty-plus with an incisive, somewhat military manner, who asked without preliminaries what I was writing my book for. I replied that I saw my task as one of asserting the role of ethics in science. Khanin retorted that this was pointless and that he could demonstrate my fundamental mistake without difficulty. As soon as we had had tea he took a pencil and paper and set about ex-

pounding his "anti-ethical" theory with the aid of graphs and calculations. It may be summarized as follows.

All processes in the universe, including the world of human
relationships, originate in terms of energy, and under the laws of
physiochemistry their tendency is to expend energy in the most
economical way. This applies also to the efforts of a society or an
individual to achieve particular ends. The whole of social activity
is dominated and predetermined by the same law, and there is no
such thing as free will in nature. Individuals, society, and the
whole of mankind act in accordance with their physical structure
and energy resources. It is thus wrong to speak of the decisive importance of free moral choice on the part of an individual or a
group. In a predetermined world or society, acts by individuals
that go against the main trend cannot be successful, as they do not
possess the necessary force to swim against the tide. The resistance of one person cannot make any difference to the upward
curve of progress. "Progress," to Khanin, means the increase of
knowledge and productivity, which together determine the future
of mankind. The activity of a moralist, a writer, or a missionary of
any kind is without purpose: morality plays no part in human
progress.

I must confess that I was disconcerted by this philosophy. It
seemed not only to cast a shadow over my personal life (and everyone believes that his life is a light to somebody) but to destroy
the meaning of the efforts and struggles of whole generations. The
only consolation was that Khanin himself thought it would be another two centuries before his ideas were understood.

How is it possible to live with the belief that physical powers
are superior to those of the soul? I ventured only a single question
to Khanin, about Joan of Arc. Patiently, with pencil in hand, he
explained to me that Joan, with all her spiritual *élan,* represented
a mere speck in the general balance of energy, a minute disturb-
ance in time and space, with no significant effect on humanity, history, or progress.

It is Khanin's teaching, then, that a man must not follow the
moral promptings of his heart, as it is a waste of energy and therefore pointless to do so. We must live our lives in such a way as to
expend as few calories as possible. True, Khanin himself is said to
argue with his superiors from time to time, for instance, about getting a certain machine constructed on schedule. A machine is a

necessary object, an element in the technical and economic advance of mankind. But as to justice, morality, and the fate of individuals, these seem to Khanin purely personal matters and certainly not worth fighting for.

Khanin's world, in which we are all robots and puppets of the Great Law of Physics, has at least the advantage that it only exists as yet in the mind of its creator and a few scientists who follow his lead. One of his admirers, Doctor of Biology Viktor Mikhailovich Inyushin, goes a good deal further, and plans to put his ideas into practical effect with the aid of up-to-date biological methods.

On my first meeting with Inyushin at Alma Ata in May 1972 he struck me as an eccentric, one of the familiar kind of researchers who are interested in nothing but their own brilliant ideas. A tall, thin man in his early thirties, his faraway look, unkempt hair, and untidy clothes marked him as a fanatic devotee of science. He was bespattered with mud, having ridden to his office through the rain on a small, noisy motorized bicycle. Looking neither to right nor left, he strode to his laboratory in a somnambulistic fashion, put on a white smock, and at once began discussing with his assistants the results of an experiment set up the previous day. His colleagues respected him as a scientist but admitted that he never went to the movies, that if anyone dragged him to the theater he went to sleep, and that the only books he read were scientific ones. In conversation with him I had the same impression of a one-track mind: he answered my questions in a gloomy, absentminded tone and appeared to be thinking of nothing but his experiments.

He had received his doctor's degree at the age of thirty and was at once put in charge of a large biophysical laboratory; soon after that he became a university professor. Perhaps in Europe or America a meteoric career of this sort is not uncommon, but in the Soviet Union it appeared miraculous, especially in a national republic. Inyushin's research was on the light radiations of living tissue, which, he had shown, could be used to measure the reserves of energy and the physiological condition of organs and tissues. A fall in the level of an organism's "vital energy" (that word again!) could be corrected by the use of red rays, which carry an especially high energy charge. On this basis Inyushin proposed to cure certain diseases by exposing the patient to a specially directed ray of pure red light. He had constructed a generator for the purpose and collected a large team of medical doctors, biochemists, physicists, physiologists, and engineers to study the effects of light on living organisms.

Inyushin's laboratory seemed to be a place full of creative enthusiasms, though the atmosphere was slightly marred by the severe, reticent manner of his assistants and the fact that they all seemed a bit frightened of their chief. Many aspects of the work proved to be secret, indicating that the laboratory had connections with the armed forces. However, on returning to Moscow I wrote an article about Inyushin and his ideas, and planned to revisit Alma Ata in autumn; he interested me as a person, and I hoped to get to know him better and include a chapter about him in my next book. In August, however, he arrived in Moscow for an international congress of biophysicists, and my wife and I invited him and three of his female assistants to spend an evening with us. It was a very hot summer that year, and the delegates were very tired. I decided not to start any serious discussion at dinner, but to give our guests a chance to relax and forget the day's toil in light conversation. At first this was successful: they seemed to feel at ease, and Inyushin himself chatted and joked pleasantly. Likewise in a joking tone, I asked him how he kept his motley team with their different scientific interests together: did not physicists and doctors of medicine, for instance, chafe at being members of a single outfit? This set him off on a monologue which I found completely unexpected. It went like this:

"When I was a student in the physiology department and had interesting ideas from time to time about biophysics, I did not set myself up against my chief, who was a Kazakh, but helped him to write his doctoral thesis. If I had said that my ideas were more important than his thesis, as in fact they were, I would have been thrown out and would have wasted ten years drifting from one laboratory to another. As it was, I used a bit of Jesuitry [Inyushin's own word] and managed to do the work that interested me while remaining in the same department. When I became a laboratory head I realized that a junior research officer must be sat on occasionally, to prevent him forgetting about the laboratory's main purpose and its overall aims. That is necessary in the interest of scientific progress."

At this point my wife and I glanced at each other, but our three lady guests showed no embarrassment at Inyushin's frankness. Meanwhile he proceeded to enlarge on what he had said.

"A personality is only valuable if it is set on a progressive course. It cannot have any independent, intrinsic value, because every individual is, first and foremost, part of a whole, namely of society. Society's chief need is progress. Hence the value of a personality depends on how far it contributes to social and technical progress and the perfection of society as a whole. Modern man, having achieved a fairly high level of material culture, is inclined to turn inward on himself and his own mental universe, with the result that we see around us increasing discord and conflicting ambitions. But the effect of progress should be to produce a harmonized society, one in which each person's aspirations combine with others' for the good of the community, like the various cells of a body. This kind of harmony is what we should plan for. We must harmonize society on a scientific basis, and for that purpose people must forgo their private tastes and ambitions for the sake of general progress."

I interrupted Inyushin to observe that this kind of harmony seemed to me no better than cruel standardization, leading to impoverishment of the spirit; that it was inimical to creative initiative, art, and science itself; that it had already been tried, the last instance but one being the Third Reich; and that the failure of such attempts was fresh in everyone's memory. But Inyushin replied, unperturbed, that previous attempts to harmonize society had failed only because they were not scientific enough. He himself had worked out a scientific plan: in the course of a few years he hoped to construct a "psychogenerator," an apparatus which could work at a distance to curb the passions of some individuals, instill courage and optimism into others, and so forth.

"If, for instance, you installed this machine in a textile factory at Ivanov employing some thousands of women, you could use it to calm the hysterical ones and cheer up those who were overworked or unhappy in their personal lives, and that, of course, would increase productivity all around. You could also use the machine to stimulate tired workers to sexual activity, if you wanted to raise the country's birthrate."

It is a host's duty to be calm and restrained whatever he may think of what a guest is saying, but at a certain point my poor wife could stand it no longer. "How glad I am," she said with a sigh of relief, "that I shan't live to see the time when your ideas come true."

"Oh, but you will," replied Inyushin firmly.

We began to wonder if he was a little unbalanced; but the three ladies seemed in no way surprised or shocked by his views. What is more, his superiors at Alma Ata seem to find them thoroughly acceptable, as they have put him in charge of administration. Indeed, his outlook is in perfect accord with the criteria of Soviet society, which he himself regards as the height of perfection. "No one can, and no one shall, overthrow the communist ideal," he declared that evening. Any minor defects were only due to the fact that society was not yet governed scientifically enough, that "the right methods" had not been found.[7] I may mention that Nikolai Dmitrievich Devyatkov, one of the most respected figures in the Academy of Sciences of the USSR—he is in charge of all secret research work on electronics—is strongly in favor of Inyushin and his experiments. He has presented Inyushin with several dozen lasers, has paid for many of his experiments, and evidently expects to profit from their results if he has not already done so.

Well, I gave up the idea of writing any more about Inyushin and his laboratory, although he invited me several times to pay another visit to Alma Ata. Only once, four years after that memorable evening, I asked some friends to whom I was writing at Alma Ata how Inyushin was getting on. The reply came that he was "in fine form."

Inyushin is in fine form, but most scientists are not too keen on his ideas, any more than on those of Dr. Khanin or Academician A. D. Aleksandrov. Run-of-the-mill scientists are suspicious of zeal and fanaticism, and their abler colleagues regard the claims and pretensions of the state as a tiresome though unavoidable accompaniment to their lives. As for self-seekers and degree-hunters, they are even less inclined to prostrate themselves before the idol of the "common cause." To them a research institute is simply a place of work; the work may be interesting or dull, rewarding or ill paid, but what has ideology to do with it? The mere mention of "service to the nation" generally provokes an ironic smile. This indifference to social issues is common to eminent and minor scientists, men and women of high moral standards and of none. People are tired to death of meetings, political study groups, and public assessments of the success of "socialist emulation."

One cause of the alienation between the state and its more creative subjects is that the increasingly ossified regime is no longer capable of accepting even what scientists are prepared to give it.

I remember an evening party in Moscow at which, among others, four economists were seated around a table. They discussed their experiences with great frankness. A doctor of sciences aged forty-five, the author of two monographs, said that prices on the home market bore no relation to production costs and therefore did not serve to regulate supply and demand. The arbitrary fixing of prices from above was destroying the country's economy, but no one was allowed to say so in print. Another expert whose work was connected with the State Planning Commission (Gosplan) said that every year the targets of the five-year plan were "adjusted" so that failures could be dressed up as victories. A third complained that neither Gosplan nor the Council of Ministers nor individual ministries would listen to anything economists told them. An official, no matter what his grade in the hierarchy, did not want to see any economic changes or improvements. Those who listened to this conversation took it all as a matter of course; only one of them, a young candidate of sciences, kept asking: "Is there really no way in which a person of ideas can make himself useful in our society?" He was not yet capable of grasping what had long been clear to his seniors, and kept trying to explain what marvelous benefits the country might enjoy if his recommendations were adopted. The others merely shook their heads skeptically.

The impossibility of getting a hearing in the right official quarter discourages many from trying to invent or discover anything new, or even publish any ideas they have. There are, however, optimists who, although they realize that the regime is one of economic and political stagnation, find it possible to console themselves with a formula on these lines: "Anyone can criticize, but I at least am getting ahead with my work. We should talk less and work more, and then everything will come right." How exactly it will come right, the optimist prefers not to think; anyway he hasn't time to think, he is busy working. The idea of honest labor as a cure for all social evils has its adherents among people aged thirty-five to forty-five who have interesting ideas and really want to pursue them, but who have as little capacity for social analysis as the cart horse Boxer in Orwell's *Animal Farm*. People of this type are not prepared to regard themselves as unpaid debtors of the state, and consequently they are out of sympathy with the most hidebound of their elders, but they have not much in common with the young generation either. I once asked a "slogger" of this kind, a geneticist and ichthyologist, whether the juniors in his laboratory shared the view that if they worked hard enough they would be protected from social and moral evils. He merely shrugged and replied sadly: "They don't want to work, and they don't know how anyway."

The slogan of intensified work has its attractions, but very often it is chiefly an excuse for not taking too close a look at social conditions. The man who says, "At least I'm getting on with the job!" is shaking off responsibility for everything that goes on outside his laboratory. There is even an idea among such people that an interest in social problems distracts a researcher and blunts his creative energy; he should be allowed to work "without having to think."

The champion of hard work who thus frees his mind from social problems at the same time casts off any thoughts of moral responsibility or morals in general. Once, when giving a talk to physicists at the Science City of Dubna, I mentioned the courage displayed by the professors and docents of Moscow University in 1911, when more than 120 of them resigned in a body in protest against police interference.

"Well, what of it?" said a member of my audience. "Do you realize what happened to the level of physics teaching in the university after that?"

Thinking he was joking, I looked around the hall, but there was not a smile to be seen. University teaching jobs were a serious matter. . . .

To work without having to think is such a convenient formula for all concerned that one might think it was invented not by scientists, but by officials of the ministries and the party's Central Committee. At all events it encases the researcher's mind in a hard, impenetrable shell.

A remarkable incident happened in March 1976 in a camp in Mordovia, southeast of Moscow. A prisoner was summoned to camp headquarters from a barracks housing Ukrainian intellectuals who were serving sentences for nationalist activities. The guards fitted him out with new clothes and sent him back to his hut. His comrades smelled a rat, and, sure enough, they discovered inside the padding of his new jacket a transistor radio no larger than a button, and some equally tiny batteries. This triumph of modern technology was, of course, fitted for the purpose of enabling the guards to know what the prisoners were talking about. Their conversations were recorded and, if they were disloyal to Moscow, further sentences could be handed out, as actually happened in some cases.

We may doubt whether the inventors of devices like these know to what use they are put; one wonders how the Mordovian incident would strike them.

I once talked about this to Lev M., a candidate of physico-mathematical sciences in Moscow, whom I met in the house of some friends. I had heard of him as a hardworking and capable physicist employed at a secret Space Medical Institute; he was about to take a doctor's degree, was said to be a decent fellow, had been an amateur boxer but owed his success to intellect rather than brawn. He turned out to be a solid-looking man of forty-two, casually dressed and with a pleasant manner. He listened carefully to the Mordovian story, thought for a while, and then answered:

"If I were to start thinking about the use that the military or anyone else put my discoveries to, I should go mad or hang myself. And if I quit my job for ethical reasons I would not only forfeit my salary but cease to exist as a human being. I love my work, my ideas, and my equipment, and I value them more than any moral abstractions."

Such were the views of the budding Dr. M., and I believe 80 percent of his contemporaries and colleagues would have said the same. These are the rank and file of Soviet science, who live on it and profess its ideals. M., it should be stressed, is a man of strict honesty in personal and domestic affairs as well as at his work, and he would not dream of committing a crime; but moral obligations of the kind in question are so "abstract" that he has no time for them. To get rid of the irksome burden once and for all, he dragged from the depths of his memory the term "abstract humanism." As a schoolboy in Stalin's day, and later in the research institute, he, like all of us, had been taught that abstract humanism was a pernicious invention of bourgeois philosophy.

Work, to him, was another matter—work was valuable for its own sake. Even this, however, is not quite a straightforward matter for him and others who are both capable and industrious.

Work does not interest them so much from the point of view of the finished product, the fact that they can enrich their country with new machines or medicines or plant varieties; it attracts them as a way of deriving personal enjoyment from scientific research, and avoiding bothersome considerations of a nonscientific kind. "So the Institute's new director is a rogue?—well, it can't be helped, I daresay he won't eat me, anyway I'm getting on with the job. They won't let me attend an international conference on my subject?—all right then, I'll manage without. So that decent fellow, the head of the next-door lab, has come to grief—well, what can I do about it? I'm up to my eyes in work anyway. . . ."

Scientists like these, capable men and perfectly honest in everyday life, compromise with the system at every turn. They are highly popular with their superiors because they work hard and because it is easy to control them by playing on their vulnerable spot, namely the fear of losing their position. For instance, Academician Solomon Giller is called in by the Latvian Central Committee and a piece of paper is thrust under his nose: "You've got to sign this letter against Israel. It's a political matter." Giller is disgusted, but he consoles himself with the thought that otherwise his valuable Institute of Organic Synthesis would be in trouble. Naturally not a word is said about this by the Central Committee, but both sides know that if he digs in his heels he can say goodbye to his job at the Institute, which he created with his own hands. So the upshot is that the Institute remains unscathed, while Giller is besmirched for the rest of his life.

Scientists who know their job and want to work hard are more victimized by the bureaucracy than anyone else—they are fair game, and often they play into the hands of officialdom. Nikolai Vavilov, the great biologist, allowed himself to be enmeshed in this way in the 1930s. The authorities knew how passionately he loved his work, and cleverly exploited the fact. "You want to look for wheat varieties in Afghanistan? By all means go there, but don't forget, while you are about it, to photograph the forts on the Indo-Afghan border" (1924–1925). "You are interested in the genetics of agricultural plants in Europe and America? All right, off you go, travel as much as you like, but be good enough to deliver some lectures praising Soviet science, Soviet agriculture, and the Soviet state. This is most important for the sake of our political prestige and to combat bourgeois propaganda" (1931–1933). Did

Vavilov feel any pangs of conscience when he complied with perfidious requests like these? As far as I know, he did not. When his friend Academician V. V. Talanov reproached him for "amoralism" in public affairs, Vavilov rebutted the charge with amazement: his trips to Afghanistan, Europe, and America had been undertaken in the purest interests of science. It was in this way that Vavilov was gradually brought to carry out the authorities' supreme behest by training up the "people's scientist" Lysenko, who was responsible for his disgrace and death. Thinking to protect the Institutes under his control, he finally destroyed them; endeavoring at all costs to pursue a scientific career and nothing else, he ended up in the Lubyanka jail.

But that, it may be said, is all ancient history—things are different now. Oh no, they are not. To the great majority of capable scientists, then as now, politics and the public interest as officially interpreted take precedence over all considerations of personal morality. Professor Buynitsky, of Leningrad University, a meteorologist and a basically decent man, once said to an assistant of his who suffered from an overdeveloped moral sense: "So you want to protest against injustice? Go ahead, then, but don't bring us into it. We are scientists and want to get on with our work. Protests, however moral, don't mix with science—so may I please have your resignation?"

Buynitsky is a veteran of Arctic expeditions, including that of the icebreaker *Sedov,* and presumably he is neither a coward nor a reactionary. He knows as well as his assistant to what moral degradation a Soviet scientist is exposed throughout his life; he recognizes that others are entitled to fight for justice, but he puts science above everything else—science, and its utility to the Soviet people.

Such is the attitude of capable and hardworking men and women; but in the seventies they are getting few and far between. Instead we have the scientific mass, who have no illusions about socialism and to whom science is only a respected and well-paid livelihood. The more overt cynics, as soon as they have defended their dissertation, hasten to organize their lives in accordance with the dictum of a Ukrainian academician, Konstantin Yatsemirsky. "The bosses are like the sun," he used to say. "Get too close to them and you're burned up; stay too far away and you'll freeze."

Not all, of course, rush for their place in the sun, but the majority seek out a position in which they are as little troubled as possible either by the claims of science or by social obligations and political control. To evade official rules, hoodwink one's superiors, and do as little work as possible has become the principal ambition of vast numbers of scientists of the junior and middle ranks. Party committees in many research institutes complain of the "decline of political consciousness" and of what they call "internal emigration." Higher circles are also alarmed at this state of affairs, and the press frequently draws attention to public apathy in regard to social duties. At closed party sessions, "internal emigration" is the subject of warnings and threats by senior lecturers and instructors. Nonetheless, it is an unstoppable process. The falsity of political study groups and philosophical seminars, of socialist emulation agreements and propaganda drives is so manifest, and the hope of putting one's own scientific ideas into practice is so small, that junior and middle-rank researchers who are not overambitious concentrate on surviving and on keeping out of sight as much as possible.

What all scientists have in common is their terror of the authorities: this applies alike to internal émigrés, to those who would lay down their lives for the workers' and peasants' state, and to the academicians, professors, and directors who run the system. Fear is the key to the mentality of a Soviet scientist. I do not know if it is true that the KGB keeps a file on every graduate of a scientific institute, but anyone who has had a higher education, including research officers of all grades, always feels that he is watched with special vigilance. There are so many open and secret prohibitions that it is impossible to foresee and remember them all. I am not speaking now of the constant fears that beset the life of an overtalented scientist whose work attracts the attention of foreign colleagues, or the tribulations of one who is a Jew, or the panic of a laboratory head who finds that he has a dissident on his staff. People in these categories live in a state of unremitting terror. But even those who are not Jews or dissidents and have nothing on their conscience cannot feel at ease, for one can never know what the authorities will choose, at a given moment, to regard as criminal and subversive.

Every summer a mathematician from Donetsk, a corresponding member of the Ukrainian Academy of Sciences, rents a villa outside Moscow next to the home of some relatives of mine. A man of excellent character, his ruling passion consists in acquiring books, on which he spends most of his salary. When I walked past the villa at any hour of the day, he was nearly always to be seen writing on the veranda. After we became acquainted he told me that he was not writing mathematical treatises, as one might have supposed, but was copying out by hand rare editions of poetical works by Akhmatova, Voloshin, Gumilev, and Tsvetayeva; in the course of one summer he had copied into notebooks no less than eight hundred pages of print. When I asked him why he did not hire a typist to copy these out-of-print books, he raised his hands in horror. Apparently the party authorities at Donetsk were extremely strict about any form of copying, and if literary works were copied onto standard-size sheets it counted as *samizdat* (the illegal private reproduction of texts). It would be dangerous to keep on one's shelves a typed copy of Anna Akhmatova's poems; the professor's students and assistants came to his house, and any of them might denounce him. And so, in the age of linotype and duplicating machines, the professor toiled over his notebooks, as this form of copying was not yet proscribed by the Donetsk regional committee. I asked him if he did not think the prohibition strange. "I never thought about it," he replied in a puzzled tone. "Yes, of course . . . I suppose so . . . But the law is the law, and I can't risk losing my job."

The professor's attitude is not hard to understand: if he once fell foul of the local party authorities he would be out of his job in no time. But others who, one might think, are protected by their high social standing are no less terrified. I listened to a broadcast from Stockholm when the Nobel prize was awarded to Academician Leonid Vitalyevich Kantorovich. At a press conference immediately afterward he was asked what he thought of the fact that his fellow prize winner Academician Sakharov had not been allowed to attend the ceremony. Kantorovich replied in an expressionless tone, and in terms evidently memorized beforehand, that official explanations had been given on the subject and that he had nothing to add. The reporter, undaunted, pressed for Kantorovich's *personal* view, whereupon the latter took fright and stammered: "The correspondent may repeat his question as often as he likes." The press representatives at Stockholm may not have fully understood the absurdity and shamefulness of this exchange, as transmitted by the interpreters, but in Russia we felt saddened

and disgraced by it, realizing that our compatriot was in the grip
of humiliating terror.

The life of a Soviet scientist is ruled by terror even when he
does his best to conceal the fact from himself and others. On a
February evening in 1976 I was visiting Dr. G., a friend of mine
since I had written about his work many years before. At about
eight o'clock he started to look at his watch, and finally said that
he would not be sleeping at home that night, as he was on duty.
The Twenty-fifth Congress of the Communist Party of the Soviet
Union was in progress, and during that period party members
were obliged to remain on call at all the research institutes, laboratories, and similar establishments in Moscow. Dr. G. knew as
well as I did that this was a pointless arrangement, yet here was
he, no longer a young man and not in the best of health, preparing
to spend the night in an empty laboratory. "Why didn't you refuse?" I asked. "Out of the question," he replied. "The party committee told us that it was an honor reserved for the most distinguished, and I accepted it as such." Saying no more, but with an
embarrassed smile, he started to pack his slippers into a small case
together with a snack wrapped in cellophane and a small pillow,
so as not to be too uncomfortable despite the absence of bedding.
I took leave of him near the subway station and watched as he
disappeared into the crowd—a stout man in spectacles and a hat,
carrying a small attaché case. It struck me then that just as he was
going off that evening, with ill-concealed vexation, to spend an uncomfortable night at the party's bidding, so he could be expected
to swallow his annoyance and contempt and carry out any other
instructions the party might give him, whatever they were and
whatever they implied. Why? Because Professor G. is a typical Soviet man, like Lev M. and the Donetsk mathematician and Professor Buynitsky of Leningrad: each of them one-millionth part of
the army of manipulated science.

A scientific officer is not only afraid of the direct representatives
of authority—ministry or party officials, the KGB, or the director
of his Institute. He is afraid, not without reason, of any colleague
who seems to be getting on well. In Soviet science, professional
success goes by favor. If Professor X's career suddenly takes an
upward turn, no one has any doubt that he enjoys protection in
high quarters, be it the ministry, the Regional or Central Committee of the party, or the KGB. The ins and outs will remain a se-

cret, of course, but meanwhile it is better to give X a wide berth, as you don't know who is behind him. This psychological atmosphere is one that favors the strong and crafty, while it enslaves the weaker brethren still further.

A recent example of a mysterious rise in favor is that of the geneticist and academician Nikolai Petrovich Dubinin. Persecuted under Khrushchev, he enjoyed the goodwill of the latter's successors and became director of the Institute of Genetics of the Academy of Sciences of the USSR; he was also regularly praised in the columns of *Pravda*. The reason for his fame did not become clear until 1973, when the party publishing concern Politizdat issued an enormous edition of his autobiography, *Perpetual Motion*. This stout volume was a shock to biologists and many others, as it completely distorted the history of Soviet biology for the past forty years. Writing in manifest opposition to Zhores Medvedyev, whose *Biology and the Cult of Personality* had been published abroad, Dubinin maintained that (a) Stalin had nothing to do with the arrest and death of Academician Vavilov; (b) there was no political motive behind the destruction of geneticists and genetics in the USSR; (c) Lysenko's opponents, the so-called formal geneticists, were one and all mistaken except for Dubinin himself, who had never fallen into any error or heresy. Dubinin went on to assure his readers in the calmest possible way that the quarrels among biologists in the thirties arose simply because the gifted Lysenko wanted to enrich Soviet agriculture with a larger number of plant varieties, while Vavilov and his followers were concerned with purely theoretical research. How it came about that so many of them lost their lives—Vavilov, Karpechenko, Govorov, Sapegin, Tulaykov, Levit, Agol, Levitsky, Koltsov, and many others—Dubinin did not say. Every page of his book, however, was full of praise for the party and the unswerving wisdom of its Central Committee.

No biologist who read Dubinin's book could doubt that it was written at the Central Committee's direct order. The leaders who, after Khrushchev's overthrow, had set about restoring the battered façade of Stalin's empire, had long needed a new version of history which would clear the party of having been the most formidable brake on the country's progress. More than once I had heard party lecturers inform scientific gatherings that the past must be forgotten and that there should be no witch-hunting, for Lysenko had had his merits after all;[8] and now Dubinin came along to reinforce these arguments.

The scientists who were most outraged by Dubinin's book were, of course, those who had themselves experienced the tragic events of the thirties, and who for decades had been abused and deprived of work and liberty. So great was their indignation that in December 1973 the Council of the All-Union Society of Geneticists and Plant Breeders protested to the Presidium of the Academy of Sciences of the USSR and to the Department of Science of the Central Committee of the CPSU against the publication of the book. Naturally they received no reply, and the matter was gradually forgotten. But there is another interesting feature. Two months earlier, in October 1973, I sent twelve letters to eminent Soviet biologists urging them to refute Dubinin's inventions and offering to help in getting their comments published. But alas, not one of them was willing to speak out openly against Dubinin. "I find it is not all that simple," replied D.L. of Leningrad, a candidate of biological sciences and a member of the Botanical Institute of the Academy of Sciences. "Any review which gave the real facts and said that an academician had falsified scientific history would stand no chance whatever of being published. . . . I have talked to many geneticists about it—they are all disgusted, they point out all sorts of examples of what a rogue the man is, they make fun of the academic 'embellishments' of his style . . . but none of them has anything effective to propose."

N.I., a doctor of agricultural sciences and a former pupil of Vavilov, now a member of the All-Union Plant-Breeding Institute of the Academy of Sciences at Leningrad, was even more alarmed. "Of course Dubinin's book is empty rubbish and a pack of lies; but what is Politizdat up to? Surely it's all a plot, they want to catch someone out with it." The only positive answer I got was from Science City at Novosibirsk, where someone wrote a long parody of Dubinin's book; but he did not sign it, and I never found out who he was.

As biologists are frightened of Dubinin, so historians are scared of B. A. Rybakov, director of the Historical Institute of the Academy of Sciences of the USSR, who has long had close links with the KGB and who manipulates history as the party tells him to. The linguists' bugbear is Fedor Petrovich Filin, corresponding member of the Academy and director of the Russian Language Institute, who allows no talented or independent scholars to remain on his staff. The most dangerous physicist, in the political sense, is thought to be Academician N. G. Basov, and in chemistry his opposite number is Yu. A. Ovchinnikov. Quite a number of other academicians are protected in the same way. The scientific small fry, when they meet these academic sharks, tremble and look for a crevice to dive into.

Independence is a source of fear, and fear leads to doublethink. Fear and hypocrisy are a contagion that affects practically the whole million-strong scientific army. When Professor G. comes home from his spell of night duty in the laboratory, he and his wife will make jokes about the idiot who thought up the idea of mounting guard by the telephone. When the mathematician who copied out Akhmatova's poetry gets home to Donetsk, at the next seminar on the theory of scientific communism he will quote with enthusiasm the Central Committee's resolution of 1946 condemning Akhmatova's ideological mistakes and the journals that published her work. For all such people, a double life is the order of the day.

And not only for them. Doublethink is due to the fact that relations between the citizen and the authorities are not governed by any law, and the result is felt by all citizens without exception. In many scientific research groups the atmosphere is one of constant tension and irritation, with violent quarrels incessantly arising over minor issues. We are told that this is because life is hard and people are tired, but what are they tired of exactly? Life is not easy anywhere in the world, but we do not hear of such outbreaks and internecine feuds in the laboratories of Britain, West Germany, or the United States. I believe the Russian quarrelsomeness is due to the effect of mental repression on people's emotional makeup. What the laboratory staff are tired of is hypocrisy and leading a double life. For years they have been afraid to talk to one another about anything that really matters, and have overlaid their personality with a thick shell of trivialities. Apart from the ordinary troubles of life, intellectuals suffer from the pain of not being allowed to express what is in their hearts and minds. Consequently the atmosphere in a scientific group is like that of an electrical storm—the lightning strikes indiscriminately, killing and maiming innocent bystanders. Many of those concerned do not understand the reason for their own irritation, or realize that they and everyone else are poisoned by the hypocrisy that pervades their lives.

As for those who do understand, what feeling does the situation arouse in them—mental disharmony, shame, despair? My many contacts with the scientific intelligentsia convinced me that awareness of doublethink scarcely troubles any of them. Not only is it the normal way of life of the scientific million, but it supplies each and every one of them with irrefutable arguments proving that things must be as they are and can never be otherwise.

This self-mesmerized state of the intelligentsia was once satirized in a sketch by Moscow University students. A man comes onto the stage and marches around saying in a depressed tone: "What can I do single-handed?" Another comes on and walks behind him saying the same thing, then a third, and so forth until there is a whole line of marchers chanting the same defeatist slogan. The authorities were not long in forbidding the sketch. It represents what may be called an elementary stage of doublethink, in which the individual blames everything on the regime and regards himself as a victim, bound hand and foot. In learned circles one finds a more sophisticated, enlightened form of the malady, relieved by intellectual humor, elegiac sadness, and a degree of self-satisfaction. Dr. G., of whom we spoke earlier, takes pleasure in the thought that he is getting on well in his career and, at the same time, has not become an utter scoundrel, since he is fully aware of what the state and his fellow beings are up to. Among his friends and relations the fact that he knows "what's up" and can make jokes and tell stories about it almost makes him appear a decent fellow. He realizes that under a different kind of regime he, as a scientist, could expect better treatment. But the regime is not going to change; he must live as best he can and somehow reconcile things as they are with what he would like them to be.

This moral attitude is a safe one for Dr. G., and it certainly does not threaten the regime either. The party and its committees at all levels are well aware that no amount of flag waving is going to restore the enthusiasm of the 1920s. If an intellectual wants to amuse or console himself with a semblance of freedom, he may do so in the privacy of his own home. Let him listen to forbidden tapes and even the BBC, let him tell his friends funny stories against the government. He won't be put in jail for those, just yet —but at his place of work there must be no nonsense, nothing but absolute, implicit obedience. Such, in the post-Khrushchev era, is the tacit agreement between the country's leaders and the ruck of Moscow intellectuals; such is the basis of the "enlightened doublethink" which some choose to regard as a portent of more liberal conditions. What they mistake for liberalization is simply a bargain, and a trivial one at that.

Returning from a party, I once wrote down for amusement what I called the "program of a friendly evening among Moscow scholars and writers in the mid-1970s." It began with disconnecting the telephone (with good reason) and playing a tape of songs by A. Galich. Then came vodka, potatoes with onion stuffing, and ice cream; then more Galich. Letters from friends who had emigrated to Israel, reminiscences of those who had gone to the United States. Galich again. Tea. Jokes about telephone-tapping, and finally a debate on whether the authorities' housing policy had a political motive—to make Moscow so big that it was that much harder for intellectuals to meet one another.

Unfortunately I do not know enough to give so exact an account of how my colleagues spend their mornings. I do know, however, that when they settle down to their typewriters or go off to their laboratories they cast off the "spirit of freedom" and do precisely what higher authority tells them to—until the next evening. Such are the rules of the game.

Actually such evenings as I have described are not wholly typical of the regimented intelligentsia. To understand the thoughts and feelings of the species of "Soviet man" who, without realizing it, lives and breathes in an atmosphere of doublethink, one must go a bit further afield and engage in longer conversations. In autumn 1975 I talked for some hours to Aleksandr Evgenyevich Shilov,[9] professor of chemistry at the Science City of Chernogolovka near Moscow, where he lives in a two-story villa and is deputy director of the Institute of Physical Chemistry. His remarks interested me, not because they were particularly original but rather because they weren't: I could have heard much the same opinions from hundreds of respected professors in Moscow, Leningrad, or Novosibirsk. In the present "liberal" era one can have any number of such conversations, provided the other person can spare the time and regards you as a fellow intellectual who will listen and not denounce him to the authorities.

There we were, then, on the evening of November 1, sitting in

armchairs beside a coffee table in the professor's living room, decorated with pictures by good Russian artists.

"Yes, you're quite right," said Shilov, "we don't have freedom
of the press and we ought to have—we need it badly. The press
and literature are a mirror of the country's life, and we don't want
a distorting mirror. . . . Collectivization . . . Yes, of course, our
agriculture's in a bad way, but can it really be true that Stalin
ruined the countryside as a matter of policy? I can't bring myself to
think so, though it looks like it. . . . Yes, drunkenness is universal
. . . the workers get slacker every year . . . everybody steals. . . .

"But, you know, everything's getting better by degrees. Not
long ago my wife and I went to two plays, *Wooden Horses* and
Fasten Your Seat Belts—at one time people who wrote such things
would have been arrested, but not now. . . . People used to
starve to death in famines, but now the government buys bread
from abroad. . . . Of course there are plenty of shortcomings,
but you surely can't deny the victories of the October Revolution—
they are beyond dispute."

The professor's wife, Alla Konstantinovna, comes in—an athletic, well-dressed woman of thirty-eight, holding a copy of
Renan's *La Vie de Jésus*.

"Listen, Alla, here's Mr. Popovsky saying uncritical [*sic!*]
things about the West."

Alla Konstantinovna clasps her hands in horror. "But how can
you? What about Chile? And Hitler? And Indonesia, where they
killed ten million communists!"

I tried to point out that Indonesia was not the West and that I
was not a Nazi sympathizer, but I could not stem the flow of indignant remarks about the West. The professor, who had just
come back from the United States and was going on a trip to
Sweden, was no less fervent than his wife.

"I wouldn't be a citizen of France, Britain, or America for anything in the world. The French killed a million Algerians—and
look at what the Americans did in Vietnam! As for the British,
their freedom's all for show. You can talk your head off in Hyde
Park, but the bourgeoisie stays in power. . . ."

I told them the story of how my literary hero Nikolai Vavilov
was hounded and starved to death in prison, and they agreed that
Lysenko was a scoundrel. I spoke of A. A. Fadeyev's suicide, and
they admitted it was a bad business. They also agreed that A. D.

Sakharov's fellow academicians were to blame in signing a letter denouncing him.

"I wouldn't have signed that kind of thing," said the professor.

"Oh yes, you would," retorted his wife. "If you'd been an academician, you'd have signed like a good boy."

Shilov looked vexed for a moment. He had twice been a candidate for corresponding membership, but had been blackballed each time—a fact of which he was almost proud.

"All the same, Sakharov had no business appealing to the West, where they hate us so."

"Well, who else could he appeal to?" replied his wife. "There was no other way of disclosing all the dirty work that goes on here."

Alla Konstantinovna is somewhat more radical than her husband. She comes from a poor family, and in her youth underwent privations that Shilov, himself an academician's son, has no conception of. The professor was willing to admit that Sakharov could not have spoken up for the truth in any other way, but take Solzhenitsyn, for instance (Shilov had read his works while abroad)—he wrote well, of course, but he was obviously soured and malicious. Dostoyevsky (yes, Shilov had read him too) would never have written in such an embittered way. While in the United States Shilov had read a speech of Solzhenitsyn's to a trade union congress, in which he said that socialism was always inhuman. "I ask you, is that true? Why, everything's getting better gradually. Writers, of course, still have a bad time of it, but just wait awhile —it'll come right, you'll see. . . ."

We talked about the party, of which Shilov is a member. This time he spoke in a weary, somewhat embarrassed tone. "If you join an organization, you have to accept its principles. Of course they can make you do things that are out of character, things you feel ashamed of. In theory one could imagine a future conflict with the party, but there hasn't been one yet."

Growing somewhat tired of my host's optimism, I permitted myself to say that it might do him good and increase his knowledge of the world if he could spend a day or two in the shoes of a Jewish intellectual or an honest literary man or even a junior scientific officer in his own Chernogolovka. "Yes, anti-Semitism does still occur with us, I'm sorry to say." But he, Shilov, would never stand for it in his own Institute. . . . Was he annoyed by what I had said? No, of course not, he replied politely; he only thought I took too gloomy a view. He agreed, however, that he perhaps saw things from too high a level and that sometimes their true shape was more visible to those lower down. . . . When I took my leave, he walked me back to my hotel (or perhaps he only wanted to exercise his dog). I noted once again his intellectual profile and the elegant correctness of his speech. He thanked me for an interesting talk. We shook hands, and parted with forced smiles. I did not feel that I wanted to see him again.

As I walked into my hotel room I remembered something else he had said: "A man needs freedom, but limited freedom. Freedom, as the saying goes, is the awareness of necessity." If he, Shilov, were in charge of a publishing house, he would certainly not be in favor of printing everything. "*Any* book? But what about sex, violence, vulgarity of all kinds?" I asked who he thought should decide what was allowed and what not. He hesitated: the Writers' Union, perhaps? "In that case," I replied, "we'd better leave things as they are—at least the censors we have now are professionals."

From one point of view it is easier for Shilov to conform to a double standard—it does not have to be reflected in his scientific works. Historians, philosophers, and literary critics are in a worse position, though they manage to find petty consolations. Dr. I., a philologist and writer on Maksim Gorky, is employed at the Institute of World Literature of the Academy of Sciences of the USSR. One day, in the coffee room of the Writers' Club, he greeted me as though we were members of a conspiracy.

"Listen, old chap," he said in an earnest, confidential tone. "Have you read my latest effort yet? You must, right away. Where I talk about the foundation of socialist realism, I've said things that would make Gorky spin around in his grave."

And off he went rejoicing at his own courage and ingenuity. In the first place he had managed, without his superiors noticing, to inject a grain of truth into the vast mass of official adulation, and secondly he could go around boasting to friends and acquaintances that he had done so. Such pride, such heroism, is frequently met with in Moscow—for every master of doublethink naturally wants to be admired, on top of everything else.

The fetters of the age bite deeper and deeper into the tender flesh of human morality. "Fiery steeds are known by their brands"

—and never, perhaps, did those brands appear so cruel to me as when I used to lecture to scientific audiences. I remember in vivid detail the three last occasions, in 1974–1975. The first was at Podolsk near Moscow, in one of the innumerable institutions known only by post office box numbers. The engineers and physicists sat in a long, dimly lighted clubroom, with rousing slogans on the walls and a kind of stage in one corner. There was no semblance of taste or comfort: it was simply a hall on the ground floor of the multistory hostel in which the scientists lived two to a room, so that they might more easily be supervised and "organized." Those who did not want to have "education" forced on them had escaped to the movies, but about a hundred had been roped in to hear what I had to say.

They struck me as an attractive lot of young men and women—tall, good-looking, athletic, some of them quite smartly dressed—and they listened attentively to my remarks. But when they began to ask questions, I found that they questioned the same points as the Chernogolovka chemists a week before. In the first place they could not accept that scientific research was not, ipso facto, morally justified. At kindergarten, at school and university they had been taught that science was a supreme boon to mankind, enabling Soviet men and women to perform marvelous exploits and providing the key to a just solution of all problems on earth and in space, today, tomorrow, and always. Surely, then, it was nonsense to say that science could ever be immoral! Did I mean that each individual scientist was responsible for the morality of his own acts? But how could they, my audience, be responsible, when there were so many superiors above them? They were small fry, after all! A person under orders could not be responsible, even for bloodshed. I had told them that a scientist was free to take any decisions he liked in the scientific field—but if everyone took the law into his own hands like that, there would be chaos. For instance, there were people like Academician Kurchatov who, for moral reasons, refused to work on the atomic bomb—could that sort of conduct be allowed? After all, it was a *political* matter. . . .

Long after I had finished lecturing they continued to press around me—tall, handsome, fresh-faced, youthful, and inquiring. Clearly they were puzzled by my strange remarks and were desperately trying to convict me of inconsistency and absurdity. They looked at each other and giggled, but somehow could not find an argument that would finish me off once and for all. Then, as I was just leaving, one of them (eureka!) exclaimed: "What about Karl Marx? Don't you remember, he said that 'being determines consciousness.' Isn't that the answer? Our conditions of life and work, all our circumstances bear the responsibility for whether we are moral or not. Our being, not ourselves! No scientist today can possibly be free, his circumstances determine what he must do. That's where you go wrong, Comrade Writer!"

The rest of the young people greeted this with childlike enthusiasm, as well they might—who would not rejoice to be relieved of all responsibility by Marx himself? How wonderful to realize that you are not alone in the world, that you do everything as part of a large group in a large state in the big, complex twentieth century, and that whatever you do or don't do is the result of "social being" and nothing else!

I was reminded of Marx's doctrine once again by some biologists in the Science City of Pushchino. There, in a cozy corner of the scientists' café, I sat drinking tea with a group of young, good-looking candidates and doctors of science who—again confounding the categories of use and morality—explained among other things that moral intransigence was less beneficial to science than a judicious compromise between the scientist, his colleagues, and higher authority. Moral intolerance had a disruptive effect on a team of scientists, while compromise—only in little things, of course!—tended to weld them together. Compromise strengthened the position of a scientist in his laboratory or institute; compromise enabled him to continue with creative work, and therefore it led to the development of science and general progress. I no longer remember the faces of those who talked to me on these lines, but I clearly remember the almost physical awareness that they thought of me as a strange, alien, alarming, and irritating phenomenon. They were angry with me—and, perhaps, a tiny bit with themselves as well.

No, we certainly did not understand one another. Looking at their indignant faces, I thought of old Professor Aleksandr Aleksandrovich Lyubishchev (1890–1972), also a biologist and, more precisely, the last encyclopedist in Russia. Not only was this marvelous man an expert on entomology, statistics, mathematics, and other "serious" subjects, but he also took an interest in philosophy, religion, moral problems, and many other matters that his contemporaries no longer wished to hear about. He wrote a great deal (though he scarcely published anything in Russia), knew many people, and, while living in the provinces, used to send his academic friends essays in letter form on various social and scientific developments. I have on my files a copy of one of these letters, written in December 1964 to Academician V. A. Engelgardt, which reads like an answer to my young critics at Pushchino. Here is an extract:

> The primary factor in the activity of a true scholar is not the desire to satisfy his own or others' needs, but the inner impulse to gratify his thirst for pure knowledge. It is the tree of true knowledge, and it alone produces the fruits that men use to satisfy their necessities. This doctrine of the primacy of pure knowledge prevailed in the incomparable days of ancient Greece; it was defended by Louis Pasteur and by K. A. Timiryazev. And the whole history of science shows that those who maintain that the satisfaction of material needs is its main purpose are simply likening themselves to a certain animal in Krylov's fables.

All very true and clear; but "what is the use of torches and spectacles if people refuse to see?" I grasped this fact one day in April 1975, and decided it was time I stopped lecturing. I had been giving a talk at the All-Union Research Institute of Equipment Design. What kind of "equipment" was made at this secret establishment became clear to me when I saw a huge placard in the lecture room announcing that the crew of *Salyut 4* were grateful to the Institute for its excellent work. There were a lot of young faces in the audience, but there was something curious about them: I did not see a single lively pair of eyes. No one laughed at jokes or seemed affected by touching incidents. Nor could I arouse any interest by talking about the moral aspects of a scientist's behavior. When I had finished, a young lady sitting not far off said: "You have painted too gloomy a picture of mass-produced twentieth-century science. We here are the scientific masses, if you like, but look at us—are we immoral?" A candidate of sciences, in his thirties, sitting at the far end of the hall underneath the congratulatory message from the cosmonauts, said in an angry tone: "Moral laws can't be the same for everyone. What is moral for one present-day society may be immoral for another.

Morality must be measured by social standards. The kind you have been preaching has something one-sided about it. You aren't by any chance trying to sell us Christian morality, are you?"

Those were the last words addressed to me by the children of manipulated science, after thirteen years of public lecturing.

CHAPTER 9

THE UNSUBMISSIVE ONES

"This is the burning question of the day: should we compromise with established ways of life . . . or openly regard them as rubbish, not even worth while as a reference point?"*

We have come to the end of our study of manipulated science. What conclusions are to be drawn from it? For those who are still in doubt I would quote a passage from the Abelevs' book, mentioned in a previous chapter:

The most valuable aspect of science consists of investigations in which basically new phenomena are observed and hitherto unknown laws of nature are discovered. . . . The history of science shows that such discoveries are usually made unexpectedly, in an unexpected place and often by the most unlikely people. The nature of the discovery, too, is generally unpredictable. . . .

This unpredictability accounts for the unique peculiarity of the organization of science, viz. that it cannot be controlled from above. It is sufficient to recall Mendel's experiments on pea plants, those of Pasteur on beer, and those of the embryologist Mechnikov on starfish. Could their results have been planned, organized, and predicted in advance? Could these problems have been solved by directives? In our time it has been shown not only by examples but by objective scientific analysis that science develops in an autonomous and statistical manner which cannot be the effect of conscious direction. Attempts, however well intentioned, to limit science and confine it to a particular channel have always hampered its progress and led to its speedy degeneration.[1]

This trenchant diagnosis is fully applicable to Soviet manipulated science, an organism smitten with ethical degradation and a cancerous administrative growth. In such cases the doctors gener-

* M. E. Saltykov-Shchedrin, *Complete Works,* Vol. 4, p. 248.

ally say, "Medicine is powerless"; ordinary people cross themselves and murmur, "He's a goner"; and the historian can only add, "Amen."

But I can hear the objection: "All right, we agree with you, manipulated science is not science in the proper sense, and the pathological degeneration of its tissues has gone a long way. But is it not too soon to bury it? Are there not some gifted scientists and efficient laboratories in Russia? You yourself in your book mention dozens of eminent names. . . ."

An organism never dies in a single moment. Hours after the heart has ceased to beat and breathing has stopped, live cells and tissues and even living systems can be detected in a patient's body. Certainly manipulated science is not completely dead, but at what price is it kept alive? The talent which manages to pursue its ideas in Soviet conditions, the group which works at major problems successfully and in a friendly spirit—these are pockets of resistance to manipulation, not working within the system but outside it. Talent has its own plans, its own way of seeking the truth, its own idea of what a laboratory should or should not contain. Independence is the basic condition of creativity, but it is also the main cause of the persecution visited on centers of insubordination. Talent, especially when combined with moral principles, is looked on as a seedbed of rebellion. Each act of creative independence is seen by Institute directors, ministries, and the Academy of Sciences as evidence of the perfidious refusal of an individual or group to conform to the established order. The result is that independence becomes a constant fight, with barriers and obstacles rising at every turn.

Suppose you want to repeat an investigation carried out in another laboratory? You're not allowed to—duplication is against the rules. Or suppose you want to leave your present field of study for one related more closely to a discovery that interests you? You can't do that either; the second field is outside the scope of your institute. Do you want to combine your efforts with those of foreign scientists? Out of the question—it's unpatriotic. Perhaps your experiments are incomplete; they need working at a bit longer? That won't do, your work is part of the Institute's plan, laid down by the Ministry as an element in the Five-year Plan of the Soviet Union. Ready or not ready, you must submit everything by the due date.

Prohibitions are constantly ringing in scientists' ears. Mediocrities are not bothered by them, but real scientists are—they fight back, or try to beat the system by guile, or fall into a state of despair. For, to a man of talent, his laboratory work is life itself, and when life is forbidden, the man dies.

The fight is not always a spectacular one, with waving of flags and shouting of battle cries. A man may simply find that he has the courage to dissociate himself from mass-produced opinions, or to reject the myth that he is only an insignificant part of a wise and mighty whole, or an innocent executor of blameless instructions. It is not so easy, even in the West, to feel conscious of oneself as a personality independent of social, party, or political prejudices; but to do so in a totalitarian state borders on heroism. In the Soviet Union, to do one's own scientific work is far more dangerous than to sit in the laboratory doing nothing.

For example, given the undisguised Great Russian chauvinism of the country's leaders, it requires unusual courage to call in question the genuineness of the *Tale of Igor's Expedition,* officially revered as a unique and brilliant product of twelfth-century Russian literature. Nonetheless, Professor Zimin has for some years been brave enough to argue publicly that it is an eighteenth-century forgery. Another Moscow professor, a believing Christian, has for years been studying the problems of Orthodoxy in secret, at the risk of ruining his career.[2]

In the provinces it is even more difficult than in Moscow to preserve one's scientific independence, but there too there are fighters for intellectual freedom. The leading Lithuanian archaeologists refused to accept the ideological line imposed by Moscow, and, to escape its influence, they resolved to study only the archaeology of Lithuania itself, to the exclusion of any links with Germany or Russia. This restriction no doubt impoverishes historical studies as a whole, but it also preserves them from being falsified; more important still, it protects Lithuanian scholars from the moral effect of subordinating their scientific personality to the dictates of ideology. The "Lithuanian version" of ethical resistance may not seem very impressive to us, but only a small number of the scientific million have had the courage to go even this far.

There have, however, been some real heroes in the battle for the scientific conscience. I. I. Puzanov (1885–1971), a professor of hydrobiology at Odessa, put up a stout fight against his persecutors. I knew him well—a thickset old man, not tall, with the brisk manner of a sea dog. He had sailed around Europe and many parts of Asia and Africa before the Revolution, and had traveled much in later years also. Students and young scientists loved him for his lively, friendly character and readiness to share his encyclopedic knowledge with one and all. In matters of scientific principle he was inflexible to the point of sternness. On the basis of his own special subject—marine vertebrates—he published an article in 1954 expressing open disagreement with the views of T. D. Lysenko, who was all-powerful at that time,[3] and, in addition, he went on teaching his students what he believed to be correct. Normally, a scholar whose views are challenged should produce fresh evidence in support of them; Lysenko remained silent, however, and Puzanov was the one who had to defend himself. One "investigating commission" after another descended on his department, and he was "inspected" from the scientific, financial, and ideological points of view, while the department's work was disorganized for months. At the end of this time Puzanov addressed to the Academic Council of Odessa University a statement which, in a contemporary's words, "ought to be read standing up, like an oath." In it he said:

> It is vain for the members of certain commissions to suppose that I can be made to renounce my views. It is not for me at my age to show weakness in matters of principle, belying my convictions so as to be sure of a quiet life and a comfortable job at the end of my days. This applies to me more especially as a professor at the university of Sechenov, Mechnikov, and Kovalevsky, men who never played fast and loose with science.[4]

This reference to great scientists of the past did not save Puzanov from harsh administrative penalties; but his spirit was unbroken, and he gained a moral victory over his persecutors.

Anyone who stands up for his own moral values in the realm of manipulated science must be prepared for hard consequences. In January 1973 members of the Moscow medical faculty accompanied Ivan Fedorovich Mikhailov to his last resting place. A man of fifty and apparently in his prime, he had been struck down by a heart attack, although nothing in his life history suggested that he would meet such an end. Born in Moscow in 1923, the son of a printing worker, he became an apprentice compositor on leaving school. In 1941 he served at the front and was seriously wounded. After the war he studied at a medical institute; he advanced rapidly and occupied several senior posts in the army medical service, including an advisership in China. At the same time he remained interested in science, especially luminescent microscopy. In Leningrad and later in Moscow he became deputy director and then director of a research institute. As director of the Mechnikov Institute of Vaccines and Serums in Moscow he proved himself to be an able man and as honest as the job allowed. On reaching the age of fifty he was awarded the Order of Lenin. In appearance he was like the average director—slow, corpulent, with a coarse, flat face of which he used to say jokingly that it "simply asked to have a brick thrown at it"; he added, disarmingly enough, that the higher-ups liked that sort of face and would be sure to make him an academician one day. He was in fact a typical favorite of the regime, distinguished from the rest only by his lively mind, a genuine interest in science, and a further quality that escaped the notice of those who appointed him to senior posts. It was that quality that proved his undoing. Details apart, the story was as follows.

The Institute of which he was director had for several years, since before his appointment, been manufacturing a complex vaccine for the armed forces. For reasons that do not concern us it turned out that the vaccine was imperfect and even dangerous, and would have to be worked on for several more years. But the Ministry of Health would not hear of this; they demanded the vaccine at once, and said that the last stage of experiments would have to be made on people. Mikhailov refused, pointing out that experiments on rabbits had produced frightful results, including deep abscesses and necroses. Backed by the Academic Council of the Institute, he declared that it was unthinkable to use the vaccine on human beings.

The officials began to threaten; Mikhailov wrote a letter to the minister. In reply a commission of inspection was appointed, its purpose being to remove the unruly director and have the vaccine approved without delay. The only way Mikhailov could have held on to his post was by signing a certificate authorizing its use on people, and this he refused to do. Some of his friends said that he himself, when in the Army, had suffered from painful and ineffective inoculations, and that he wished to protect soldiers from even worse experiences. But his true motive was a deeper one. People close to him knew that, coarse and commonplace as he might seem to be, Mikhailov at heart was not only a doctor but a humanitarian, and the command to use a harmful vaccine on human beings was in his eyes immoral. A respecter of the Hippocratic oath, he regarded the pressure to which he was subjected as a crime against the moral law. He suffered a first heart attack, the result of grave moral strain, on the eve of a decisive session at the Institute at which he knew the officials would get their own way. The second heart attack came when he learned, after his dismissal, that the vaccine had in fact been used on people.[5] Addressing a crowd of friends and colleagues at Mikhailov's funeral, Professor Litinsky said: "We know what caused Ivan Fedorovich's death, and we know who the guilty ones are. He died as a hero."

Mikhailov's action was indeed heroic, the more so because it was performed by a director. Thousands of men in a similar position comply with orders by signing the most inhuman documents every day. Mikhailov found the strength to avoid being sucked into a moral quagmire. His death brings to mind Albert Schweitzer's words about Einstein: "Einstein died of the realization that he was responsible for exposing mankind to the danger of atomic war."[6] We are not comparing Mikhailov's talents with the genius of Einstein—it is not a question here of talent, but of the scientific conscience and a sense of responsibility, which the Russian microbiologist showed as truly as the American physicist, and at much greater risk to himself. . . .

As I have already said, the number of recalcitrants among the scientific million is not large; but whatever the personal fate of each one of them, the lesson of their fight is that a scientist can function as such in the Soviet Union only by a moral struggle and by refusing to be coerced: there is no other way. If he adopts any other course, he cannot be a true scientist, for science means nothing but independence of mind. This idea, taken for granted in the West, has only made its appearance in the Soviet Union in the past twenty years or so. Three years after Stalin's death, when the "thaw" was just beginning, Professor Aleksandr Aleksandrovich Lyubishchev of the Ulyanovsk Pedagogical Institute summarized his advice to young people embarking on an academic career in two words: "Be independent." Lyubishchev—a biologist and philosopher, then aged sixty-five—went on to tell his audience that they must be independent (a) of those around them—"You are your own supreme judge," (b) of circumstances, (c) of narrow specialization, and (d) of all kinds of dogma. The last point he expanded as follows: "If one can draw any indisputable conclusion from the history of civilization, it is that even the most progressive teaching, once it becomes a dogma immune from criticism, acts as a brake on social and scientific progress."[7] Lyubishchev's advice was contained in an article written for a young people's newspaper; needless to say, it was not printed, but it circulated for a long time in *samizdat* and helped to arouse many from a state of ethical hibernation.

Nowadays, in various parts of the country and in various research institutes, one comes across people whose life-style is based on the moral imperative and on release from spiritual enslavement. Despite the very real risk of being deprived of a livelihood, they do their best to live honestly in the professional sphere. Fifteen years after Lyubishchev, the Abelevs in their turn formulated the demand for high moral standards:

> One of the chief motives that impel a scientist to behave ethically is a certain inner necessity, a sense of duty, the urge to preserve his internal balance and integrity. We know little about our inner world, but we do know that it is indivisible. . . . A scientist is productive only if he possesses internal freedom, follows his own interests, and trusts his own opinion; he must live emotionally in the world of scientific concepts and feel that he is part of the flesh and blood of science. . . . He well knows that if he departs from ethics he undermines the very basis of science, doing it a mortal injury that all his research activity cannot make up for. Thus his own scientific work becomes a mockery. If he is false to his own scientific conscience, he kills the researcher in himself. That is why, as a mere matter of self-preservation, all "prudent" calculation to the contrary, he should strive obstinately, mulishly even, to retain his ethical purity. We do this above all for our own sake, not to help anyone else and not for the sake of any specific result. We do it with annoyance and vexation at times, but the real reason is that we cannot do otherwise.[8]

Such is the credo of a handful of "rebels" interspersed among the docile million. The words of the Abelevs may also serve as an ethical portrait of the laboratory under Garri Abelev's charge.

An individual scientist or a whole laboratory that recognizes the moral imperative is liable to be persecuted, not only because the Soviet state apparatus and society as a whole do not regard ethical norms as binding. Those who think for themselves are at risk primarily because their ethical protest is interpreted as a political one. There is no neutral ground in Russia between the intelligentsia and the authorities: it is not possible for teachers, doctors, writers, engineers, or scientists to coexist with the powers that be, preserving their own ideas and convictions. When the poet Yosif Brodsky was asked in America why he had left the Soviet Union, he replied: "In my country a citizen must either be a slave or an enemy. As I was neither, the authorities didn't know what to do with me, so they threw me out." This half-joking reply is an accurate picture of the official Soviet attitude. The state apparatus either sucks you in and makes you serve its ends, or it throws you out onto the barricades; there is no third possibility.

Any and every ethical protest is treated by the authorities as a political crime and judged accordingly. How exactly the KGB will treat a scientist who offends in this way cannot be predicted in advance. The boundaries between moral protest, political disloyalty, and criminality are so blurred that a citizen never knows precisely where he stands. At times perfectly neutral and law-abiding scientists have found themselves cast as political fighters when they had no intention of being such. The eminent surgeon V. F. Voyno-Yasenetsky (1877–1961) spent twelve years as a prisoner and deportee for the sole crime of trying to combine faith and science despite the regime's intolerance of religion. At a party conference in the late sixties Professor Meshalova, director of a Moscow research institute, spoke of the difficulties of laboratory assistants and junior scientific officers without a higher degree. As they could not live on their meager pay, they had to go in for a scientific career even if they were not really interested in research. Meshalova was a perfectly loyal citizen, but in the heat of the moment she went on to say that her conscience was offended by the inflated salaries of secretaries of the party's Central Committee. She was at once expelled from the party and dismissed from her job; true, she was not arrested, but the political charges leveled against her came very close to involving penal sanctions.

The vagueness of the law and legislation, the whole purpose of which is to keep the population on tenterhooks, sometimes paralyzes the most natural human feelings. In March 1976 I attended the civil funeral of the biophysicist Grigori Podyapolsky, a friend of Academician Sakharov's and a member of the Committee for the Defense of the Rights of Man. I invited a woman biologist who had known him well to go with me. "Will I get into trouble?" she asked, and, to be on the safe side, decided not to go.

The uncertainty of the law enables the authorities to exert pressure on ethically minded scientists by meddling in the most intimate details of their lives. In 1975 Professor Valentin Sergeyevich Kirpichnikov, an elderly geneticist in Leningrad, learned that his favorite pupil, Sergei Kovalev,[9] a candidate of biological sciences, was about to be put on trial at Vilnyus. Moved by sympathy for the young man, he went to Vilnyus in the hope of attending the trial, but found that it was political and therefore secret. However, at the courtroom he was shown a telegram addressed to the registrar of the Supreme Court of the Lithuanian SSR, which read: "On account urgent work please ensure immediate return of V. S. Kirpichnikov to Institute of Cytology." The telegram was signed by Afanasy Troshin (born 1912), the director of the Institute and a corresponding member of the USSR Academy of Sciences. Troshin well knew that there was no urgent work and that Kirpichnikov, a consultant at the Institute, was not obliged to be there every day, but he was acting under KGB orders. The authorities wanted to keep Kovalev's trial a secret and therefore to restrict the number of his friends who showed up at the courtroom. The object of the telegram was to keep Kirpichnikov out of Vilnyus by administrative means for the duration of the trial. When he returned to the Institute, he was reprimanded by the director. Thus a teacher's natural sympathy for his victimized pupil was regarded as a political demonstration and punished as such.

In the same way a large number of fighters against official immorality have been branded as political criminals. Among those recently released from prison or exile are the physicist A. Tverdokhlebov and the astrophysicist K. Lyubarsky. Still in jail are the physicist A. Shcharansky, the geneticist S. Kovalev, the philologist G. Superfin, the physicist Yu. Orlov, the historian V. Moroz, and many others. All will remember how the cybernetics expert General P. Grigorenko was for a long time incarcerated in a psychiatric hospital. Émigrés who have left the USSR largely for moral reasons include the physicists V. Turchin, A. Voronel, and M. Azbel, the cyberneticist M. Agursky, the mathematicians L.

Plyushch and A. Volpin-Yesenin, the biologists S. Myuge and R. Berg, the philologists I. Melchuk and Tatyana Khodorovich, and the literary historian A. Yakobson. Many scholars and scientists are still leaving the country as a result of the insoluble ethic conflict between them and the authorities.

Certain Western broadcasts in Russian give the impression that there is in the USSR a coherent organization of dissidents, a social movement that comes into daily conflict with the authorities. Nothing like this exists, nor would it be possible in Soviet conditions. Those who are called dissidents in Western broadcasts have no single program or common set of aims. When I conversed with scientists and others who signed letters in defense of prisoners, recorded violations of the Helsinki agreement, etc., I noticed that their views diverged to a remarkable extent. This is not an accident, nor is it just due to Russian lack of organization and coordination. The people in question are no less determined to preserve their independence in public affairs than in the scientific field. When I asked Professor Yuri Orlov in December 1976 why he took the risk of being deprived of freedom and tranquillity, he replied that his object was to compel the Soviet authorities to observe the laws and to liberalize political institutions and social conditions. He believed that public protests had more than once caused the regime to make concessions in particular cases. The bureaucracy is afraid of publicity, especially of an international outcry, and sometimes lets a victim escape from its jaws so as to appear less tyrannical than it really is.

General Grigorenko, now an émigré, defined his aims more narrowly in terms of direct and immediate results: the release of illegally arrested prisoners, or permission for the deported Crimean Tatars to return to their homeland. He quotes Lenin and the Constitution to Soviet officials in the hope that they will be ashamed of their atrocities and set free the latest victim. He wants to help individuals at once, independently of the social situation.

Then there is a third line of resistance, represented by those whose efforts at reform are based on religious feeling. The linguist Tatyana Sergeyevna Khodorovich says: "To people languishing in prison cells and camp barracks, the most frightful thing is the thought that people outside have forgotten them. It is our duty to assure these unfortunates that they are not alone. We tell the whole world about them, we encourage them as best we can, and moral support of this kind can be more important to them than food parcels."

It might be thought that "political" campaigners like Yu. Orlov, pragmatists like Grigorenko, and Christians like Khodorovich have nothing in common except that they abhor the cruelties of the totalitarian regime. But they are also united by the fact that they are all striving for the triumph of genuine moral values. To them a citizen is not a pawn of state interests but an individual who should be able to choose where he lives, what work he does, and what he believes in. The social conflict in the USSR is between men and women with a conscience and the essentially immoral party-state apparatus.

According to Academician L. A. Artsimovich (1909–1975), "Science lies in the hands of the Soviet state and is sheltered by their protecting warmth." The reader of this book will hardly see the matter in so idyllic a light. It is in fact perfectly clear to everyone concerned that science is gripped in the viselike fist of the party-state tyranny: defenders of the regime are proud of this state of affairs, while liberals and radicals do their best to combat it. But no Soviet scholar or scientist is in a position to separate his creative activity from the overmastering authority of state interests, except for one group—religious believers.

Fifteen or twenty years ago, the expression "a scientist and a believer" would have seemed to me a contradiction. Like all my generation, I had been taught from my earliest years that science had refuted religion and was incompatible with faith. We had heard, of course, that Newton and Faraday, Pasteur and Pirogov, were believers, but that was all in the old days—science had not advanced so far, they couldn't be expected to understand. True, Pavlov, a Soviet scientist of the first rank, had also been a believer, but this was put down to senility or family tradition (his father was a priest). In the late fifties, however, people began to display religious feelings which they had till then been forced to conceal. Among the men of science who became known as believing Christians were the anatomist A. A. Abrikosov, the geochemist V. I. Vernadsky, the orientalist N. I. Konrad, the ophthalmologist V. P. Filatov, and many professors, notably the surgeon S. S. Yudin and the astronomer Kozyrev. At the same time it became known to us that religious faith was not extinct among Western intellectuals in the age of the scientific revolution.[10] This fact was a revelation to people living behind the Iron Curtain.

Also around that time, young academics in Moscow and Leningrad began to take a keen interest in church art and architecture, icons, and liturgical music. The intellectual public showed concern over the destruction of churches and talked about preserving the monuments of antiquity. A further stage of religious awakening came in the mid-sixties with the appearance in Russia of books by Berdyaev, Lossky, Frank, and S. Bulgakov. The *Journal of the Russian Christian Movement,* published outside the country, began to be seen in Moscow apartments. Everything in such books and journals was new to us—new, but not strange. Young mathematicians, physicists, and biologists began to debate such subjects as Christianity and culture, Christianity and morality, Christianity and science. A Russian translation of Pierre Teilhard de Chardin's *Le Phénomène humain* (1955), published in 1965, added fuel to the growing interest in Christian philosophy. True, this interest at times took on the appearance of a fashionable game. Lecturers who were not believers took to decorating their rooms with icons, while students sported amulets and wore crosses around their necks. For some time serious people dismissed the vogue for icons and imported religious books, predicting that it would be short-lived. A certain writer, highly thought of by the intelligentsia, depicted the character of a lady who combined earthly preoccupations with fashionable chatter about Berdyaev, at the same time speculating in icons. There were indeed quite a few women of this sort on the fringes of the literary and intellectual world at that time. But years passed, and the fashion for Christian books, religious arguments, and ideas showed no sign of dying out. Casual discussions became more and more thoughtful and specific. After one of these I wrote in my diary in February 1973:

> F.V., a candidate of physico-mathematical sciences, says that our backwardness in the natural sciences is not only due to witch-hunting, the low quality of cadres and the fact that we are cut off from world science. Our main trouble is the lack of an integrated world-view. Marxism is a lifeless accumulation of texts which does not satisfy the soul or explain the real world of today. Few of our young people know anything about Christianity, the European world-view, with its integral vision of man and its clear-cut ethical system. The

spiritual disunity of the world prevents our understanding even physics properly.

By the early seventies many young intellectuals felt a keen desire to learn about Christianity and relate their professional activities to the Christian world-view. Groups for the study of the Old and New Testaments were formed secretly in Moscow. Here is a note I made describing one of their sessions:

Sunday morning, near Moscow. We walked a long distance, past monotonous concrete barracks of the Khrushchev era, and came to an oasis—a tiny modern apartment crammed with books and papers, in which people were discussing the Sermon on the Mount. The host was nearly eighty years old, a philosopher and historian who had spent twenty years in camps. A frightened man. You can't just go and see him, you have to be introduced by one of his friends. I went with a young mathematician who has belonged to the seminar for over a year. The old man, wearing a shabby woolen jacket and felt boots, sat on a rickety divan; his audience, about half a dozen men and women aged twenty-five to thirty, sat around a table covered with threadbare oilcloth. Besides the mathematician there was a psychologist and a meteorologist; I don't know the profession of the others. They all looked serious and concentrated, and most of them took notes. The last two sessions had been about Judaism and the Cabala. The old man held an exercise book from which he read, making emphatic gestures with his other hand, which was white and muscular. He read slowly, with long pauses; more like a preacher than a lecturer. He tended to outbursts of rhetoric, and his monologue was almost impossible to interrupt. I would not say that his remarks struck me as especially ingenious or original, but he gave an impression of sincere faith and genuine fervor, and it was clearly his faith that fascinated the audience. When the session was over another group turned up—apparently also academics, though it is not easy to know who is who: they don't introduce themselves to one another, or even give the old man their telephone numbers. . . .

Despite all the fear and secrecy, people have the courage to read forbidden books and make notes of clandestine lectures. The history and philosophy of religion is a favorite topic of young Russian intellectuals. Slowly but surely, the number of study groups increases: after Moscow the provinces, after Christian groups Jewish ones. For the first time in half a century a native literature on the philosophy of religion is springing up on Russian soil. A mathematician who is a corresponding member of the Academy of Sciences writes a speech, widely circulated in *samizdat,* on the theme that mathematics by its very existence bears witness to God. A linguist and psychologist submits a work to the Institute of Information arguing that the world can only be understood on the assumption that it possesses an internal order, since otherwise we would not be justified in extrapolating from previous observations or expecting situations to repeat themselves. Scientific experience shows that the world does in fact possess an internal order, an intelligible unity, and a logical principle of development; and, the writer argues, this illustrates the need to combine scientific thinking and religious faith. So far not many works of this sort have come within the Soviet reader's purview, and the collection of essays entitled *From Under the Rubble* was greeted with all the more enthusiasm. Three of the best contributions to it are by Borisov, Agursky, and Shafarevich.

Talking at various times to believing scientists, I noticed that their approach to religion varies widely: some believe in the Church, others in Christ as God and man, others in Christ as man only. I have also come across some Zen Buddhists. But the commonest attitude among scientists is a pantheistic one, or "pretheistic," as one believer of my acquaintance called it.

The seventies introduced yet a further element, as many young intellectuals including scientists became practicing and believing members of the Orthodox Church. Doctors and candidates of sciences, senior and junior researchers, began to attend services, receive Communion, marry in church, and have their children baptized. What proportion of the scientific million have been touched by faith to this extent? The pattern that seemed to emerge from my questioning of Moscow and Leningrad researchers was that converts were mostly aged between twenty-five and thirty, while those between forty and sixty were usually indifferent to religion. This clearly refutes what used to be the widespread idea that religion is a defensive mechanism of old people who are afraid of death. Another point I noticed was that the converts were usually on a higher scientific level than the average. As Francis Bacon said: "A little philosophy inclineth man's mind to atheism, but depth in philosophy bringeth men's minds about to religion." There is a definite correlation between scientific ability and a sense of religion. Scarcely any of the converts I met were of the obtuse, untalented kind; religion does not attract scientific bureaucrats, who, as we know, are not remarkable for brains or learning. Many gifted young people, on the other hand, were full of religious enthusiasm. In groups of young mathematicians, physicists, chemists, and biologists, I observed a curious kind of "metastable" situation in which a single focus of activity would suffice to crystallize a whole gamut of religious feeling and aspiration. Sometimes the crystallizing agent would be a popular professor or lecturer; a cell of three or four people would then form, and would rapidly extend more and more widely.

Most of the converts find their faith a great source of joy, but life adds a dose of bitterness to it. An intellectual who is known to be a believer can expect all sorts of trouble in his professional life. Some antireligious drives are the work of "amateurs," but others are remorselessly planned by the authorities. For instance, teachers in places of higher learning are required by statute not only to teach their subject but to inculcate a spirit of loyalty to the party and to Soviet ideology. This rule is invoked by the administration whenever it wishes to dismiss a lecturer who is also a believer. If the victim is not a lecturer but a pure researcher, he can be got rid of by means of the public vacancy-filling examination, which a believer would not be allowed to take. If a student is seen going to church, the procedure is even simpler: he is expelled from college at the instance of the Komsomol organization. Of late, however, the authorities have decided to take a more radical and comprehensive line. A new decree on the Awarding of Higher Degrees and Academic Titles states (Chapter 3, §24): "Degrees may be granted to persons who possess thorough professional knowledge and attainments in a particular branch of science, together with a broad scientific and cultural outlook; have a command of Marxist-Leninist theory; have a good record of scientific, productive, and social work; conform to standards of Communist morality; and are guided in their actions by the principles of Soviet patriotism and proletarian internationalism."[11] This means that the party committees in research institutes, their Academic Councils, and the Higher Examinations Board can prevent any unsuitable person from taking a degree or strip him of a degree once he has taken it; for there is no difficulty whatever in declaring that the behavior of a dissident or a religious believer is unpatriotic, contrary to Communist morality, or at variance with proletarian internationalism.

A believer faces serious problems in his private life as well. In the first place he must consider carefully what to tell his children. If he brings them up as believers, it will mean, for a start, that they will not be eligible for higher education. Most of the academics I know have refused to compromise in this matter, but there are still a number who have not the courage to ruin their children's lives. Not long ago a friend, Doctor of Philosophy A.R., said to me: "I do not dare tell my children the truth about my religion. I am a Christian, I believe in Christ, but I cannot possibly disclose the fact to my sons and involve them in a conflict with society. My life is complicated enough as it is. Nowadays there's a strong movement among young intellectuals toward religion and the spiritual life. I can only hope my boys will take that road of their own accord and become Christians too."

An intellectual who has become a Christian can also expect to find himself in conflict with people who recently were of one mind with him in public matters. It is not uncommon for radicals to upbraid a convert on the ground that he has become passive and tolerant of social evil. I know of cases where such charges of opportunism have led to the breakup of a long-standing friendship between honest and decent people. It must be remembered that in the sixties believers and unbelievers fought together for freedom of conscience and free speech, but since then the position has changed: believers have reverted to the traditional attitude of demanding more of themselves than of others, and not expecting too much from the powers that be. When I asked an economist, a religious believer, what he advised me to say to lecture audiences about their moral duty in a world of lawlessness and terror, he replied: "Tell them more or less what it says in the Gospel. 'It must needs be that offences come'; there will always be evil in the world, but it is a man's business to see that the evil does not come through him." Many Christian intellectuals would endorse this view; but it does not follow that they have debarred themselves from resisting the evil that besets them in their places of work. I know several instances, in fact, in which believers have opposed a firm "no" to the temptations of science-manipulators.

For example, in July 1974 the Voice of America broadcast an appeal by American biologists for a worldwide cessation of experiments in gene manipulation. They spoke of the dangers that might follow from the transplanting of genes and the creation of organisms with previously unknown qualities, and proposed that an international conference be called to impose a ban on experiments. It became known that the American Academy of Sciences supported this initiative and that a moratorium had been imposed on a certain range of experiments for humanitarian reasons. It was pointed out that these experiments might lead to the formation of highly lethal bacterial cultures, which in the hands of an aggressor could be more dangerous than the atom bomb.

Two months later I heard the sequel to this proposal, not from the U.S. but from the Soviet side. In September 1974 Academician Ovchinnikov, whom the reader has met before, summoned a group of biologists to the Presidium of the Academy. Those invited were not doctors and senior academicians but included many young candidates of science and juniors with only a first degree. Without wasting words the vice-president informed them that they were confronted by a major political challenge. The Americans had stopped work on a superpowerful bacteriological weapon, and it was for Soviet scientists to make the most of this opportunity. A group of the ablest and most energetic specialists in various fields was being formed for the purpose—who would volunteer? It would be a storm troop, a force of marines, a biological commando, a crack division! All the money and equipment they wanted! No need for a higher degree! Even Jews were eligible! The one and only objective was to come up with a supervirulent type of virus or pathogenic microbe, and gene engineering was an excellent method to this end. The elite who volunteered to conduct the experiments need have no concern about their future: degrees, state prizes, and decorations would be theirs for the asking.

The young people's eyes sparkled: this was not just anybody talking, but Ovchinnikov himself, and what a career *he* had already carved out at forty years of age! They would follow him to the world's end! Only one of those present expressed any doubt. Vyacheslav G., a candidate of biological sciences, ventured to say that if two men, not especially well disposed toward each other, lived in a wooden house and one of them was on the point of inventing matches, the best thing the other could do would be to make a fire extinguisher; otherwise they might both come to grief. In other words, Vyacheslav G. (whom we met in Chapter 3) hinted that if the authorities were determined to go in for gene engineering they might do well to produce superpowerful vaccines and prophylactics to protect the Soviet people in the event of bacteriological war being forced upon them.

But Ovchinnikov would not have this. No one "at the top" was interested in extinguishers—what they wanted was matches, and the bigger the better. . . . So the argument remained unresolved: there was no common ground between the godless technocrat and careerist and the believing Christian who did not wish to be the man "by whom the offence cometh." We may be sure, however, that Ovchinnikov got his secret laboratory, if not a whole Academy institute devoted to gene manipulation and bacteriological warfare. No doubt, too, he was able to recruit some quite talented young people. Fortunately, however, not everyone is prepared to sell his soul to the devil. At the beginning of 1976 two of the ablest members of the Protein Institute of the Academy of Sciences refused to take part in the gene engineering project. Aged thirty and thirty-five, these two specialists informed their laboratory chief that their refusal was based on moral grounds. They were, as it happens, not Christians but Jews. The essential thing, however, is that they were men of faith, and that faith stands for love and not murder.

In order to get as complete an idea as possible of what Moscow and Provincial scientists thought about the relation between faith and science, I prepared a hundred copies of an anonymous questionnaire which friends of mine circulated to their colleagues. It consisted of three questions:

1. Are religious faith and scientific work compatible in present-day conditions?

2. How widespread, in your opinion, are religious attitudes among Soviet scientists?

3. What effect does a scientist's faith have on his work?

The questionnaire was given to scientists in Moscow, Leningrad, Frunze, Krasnodar, Tallinn, and one of the Science Cities near Moscow. No distinction was made as to age, sex, or position; only known informers were excluded. Out of the hundred recipients, fifty-nine chose not to reply; some of these remarked that they had no time to bother with such "idiotic" questions. What were the views of the remaining forty-one?

As to the compatibility of faith and science, five answered with a definite no; two replied "Probably," while thirty-four answered yes. The following were the most interesting answers.

Doctor of philosophy, male, aged forty-three: "Yes. Science and religion diverged from a single point, as if setting out in different directions around a circle. They reached their furthest separation in the nineteenth century and the beginning of the twentieth, and now they are coming together again."

Candidate of physical and mathematical sciences, instructor and researcher in physics, male, aged thirty-six: "Yes. Religious feelings and science are compatible in two ways. They can (a) be noncontradictory and support each other or (b) produce a dual consciousness. The first is the ideal we must strive for, the second is what I often see around me. This is not a purely Soviet problem, and not confined to science either: it is a consequence of human sinfulness. I personally see no sense in pursuing science without faith."

Laboratory assistant with higher education (physics and chemistry), male, aged twenty-seven: "Certainly they are compatible, apart from political factors which sometimes make it necessary to choose between loyalty to the Church and one's desire to work in a scientific research institute."

Doctor of medical sciences, experimenter, male, aged forty-two, Jewish believer: "Yes. For Nature and God are the same thing, existing or created in perfect wisdom and harmony. Science is about Nature and religion is about God, which is essentially one and the same, though the two seek knowledge by different means."

Candidate of physical and mathematical sciences, logician and psychologist, male, aged forty-nine: "Lord, help my folly, that I may serve Thee by my reason."

As to the prevalence of religious feelings, the forty-two answers differed widely. Ten were "Don't knows," while twelve thought the number of believers in the Soviet Union was small or extremely small. One replied "Zero percent in the social sciences" (this is not correct). Nineteen others gave widely differing percentages for the scientific million as a whole: 0.3 percent, 0.7 percent, between 1 and 2 percent (fifteen respondents), between 5 and 10 percent (three respondents), 15 percent (one respondent). Here are some of the longer answers:

Candidate of physical and mathematical sciences, male, aged twenty-nine: "Religious questing is more common among the young. People aged around thirty show the highest degree of religious feeling." (Many of those who replied made the same observation about thirty-year-olds.)

Senior laboratory assistant, biologist, female, aged thirty: "I do not know any Soviet scientists who are believers."

Laboratory assistant, physics and chemistry, male, aged twenty-seven: "It is hard to give a percentage: many keep their religion a close secret."

Doctor of medical sciences, pharmacologist, male, aged forty-six: "We must distinguish 'scientists' from 'men of science' [using the English terms]. There are many more believers among the former than among the latter. Apart from that, many people nowadays play at religiosity because it enhances their prestige in intellectual circles. Not more than about 3 percent are genuinely religious."

Candidate of philological sciences, linguist, male, aged thirty-seven, believing Muslim: "According to atheistic sociologists, religious feelings are widespread among the uneducated and are somewhat on the increase among brain workers. In my circle of acquaintance they are usually due to a nonconformist attitude and an interest in traditional culture."

As to the effect of religious belief on creative work, again many different views were expressed. Here are some of the most striking:

Doctor of biological sciences, immunologist, male, aged forty-eight: "True religious feeling is as an exalted spiritual phenomenon which ennobles, uplifts, and spiritualizes scientific work. It gives the scientist the sense of a high spiritual duty imposed on him from above."

Junior research officer without a higher degree, biologist, male, aged thirty: "Fundamental discoveries can only be achieved through religion, that is to say through a profound belief in truth. Religion is clearly the same as creative ability."

Candidate of physical and mathematical sciences, male, aged forty-two: "Creative ability is given to few and appears supernatural."

Graduate student of philosophy, male, aged twenty-nine: "The more creativity, the more faith."

Doctor of medical sciences, experimenter, female, aged thirty-nine: "Religion has a favorable effect on scientific creativity, even though it is based on an idealistic conception of the universe. I am of this opinion because moral and technical progress requires the presence of ideas and conceptions that transcend the moral possibilities of society as a whole."

Doctor of philosophical sciences, male, aged forty-three: "I agree with the Strugatsky brothers, who wrote in *The Inhabited Island* that 'conscience determines the scientist's aim, while science, thought, and reason provide him with the means of achieving it. Religion prevents the mind from choosing evil aims.'"

Doctor of medical sciences, male, aged forty-six: "Religious feeling should, first and foremost, improve relations between scientists working in a group."

Eight of the respondents thought religion had no effect on scientific creativity and professional relations, while two thought it was a hindrance in this respect. Finally, here are two different views that deserve to be recorded:

Junior scientific officer without a higher degree, mathematician, male, aged twenty-nine: "Science has detached itself from faith."

Candidate of physical and mathematical sciences, experimenter in physics, male, aged thirty-six: "Sometimes, and in Soviet conditions quite frequently, science appears completely detached from the believer's spiritual life—it is simply a craft as far as he is concerned, a way of earning his living. The scientist's spiritual life is concentrated in his faith, and he begins to find science a burden. This conflict is due to our social order and the fact that believers are persecuted."

Sociologists may consider that my questionnaire was organized and analyzed unprofessionally. I am aware of its shortcomings, but it does show clearly that religion is gaining ground among the scientific million, whereas only a short time ago unbelief was considered essential to "scientific objectivity" and faith was regarded as an antiquarian survival. Like all other people, scientists who believe do so in different ways. But the gifts of faith are beneficial to all alike: they help some to behave better to their colleagues, they inspire others to creative research, and they restrain still others from working in fields in which evil manifests itself with least concealment. It is interesting that all this should be perceived most clearly by the generation aged thirty—the age of Christ at the fulfillment of his mission, and the age at which they themselves are entering on their most productive period.

This book contains about 280 pages, some 25 of which are devoted to those scientists who reject official control. This corresponds closely enough to the ratio in real life between the submissive million and the minority of truly creative spirits. The percentage of those who refuse to be manipulated is rising slowly, very slowly; it is hard to imagine that in the next decade or so the varied company of gifted scientists who believe and who stand up for their principles will even double in number. Their increase is restrained not only by official terror but also by the dull, obstinate hostility which the rank and file of manipulated scientists feel toward any colleague who is not exactly like them. The million are as a rule spiritually closer to the party bureaucracy than to the rebellious intellectual elite. But the latter refuse to be extinguished, and it is they who will rescue Soviet science from final disintegration. As long as they stand firm, there is hope that manipulated science may become science in the true sense of the term.

Moscow, April–September 1976

NOTES

NOTES TO CHAPTER 1

1. A start in this direction has already been made. I am grateful to an unknown author who, in October 1975, published in *samizdat* [illegally circulated manuscript. –Trans.] an article entitled "Concerning a Certain Jubilee Publication," from which I have taken some facts concerning the history of the Soviet Academy of Sciences.

2. *Bulletin of the Russian Academy of Sciences*, 1917, No. 11.

3. B. V. Babkin, *Pavlov: a Biography* (Chicago, 1951), p. 104.

4. *Bulletin of the Academy of Sciences*, 1932, No. 11.

5. N. I. Vavilov, preface to V. E. Regel, *Grain Crops in Russia*. Professor Regel died of spotted fever while traveling from Petrograd to his home at Vyatka.

6. From *Gorky and Science*, (Moscow: Nauka, 1964).

7. Archives of the All-Union Institute of Plant Breeding.

8. Ibid., letter to Professor Zaytsev.

9. *Bulletin of the Russian Academy of Sciences*, 1918, No. 7.

10. In August 1919 Frenkel wrote to his mother from prison: "I am not at all bored, and am reading quite a lot. . . . I have started on my article and am now waiting for the text of my lectures, which they haven't given me yet. I play a bit of chess." From V. Ya. Frenkel, *Yakov Ilyich Frenkel* (Moscow, 1966), p. 73.

11. Central Party Archives, Institute of Marxism-Leninism, collection 461, file 31286, folios 1, 2, 3.

12. Lenin, *Collected Works* (in English), edited by Lawrence and Wishart, (London, 1965 ff.); Vol. 44, 1970, p. 284: letter from Petrograd dated September 15, 1919.

13. Lenin, *Collected Works* (in Russian), 5th ed., Moscow, Vol. 51, p. 52: letter dated September 18, 1919. (Not in English ed.)

14. Lenin, *Collected Works* (London, 1965), Vol. 29, pp. 179–80.

15. Ibid., pp. 228–29.

16. And, in order to keep them in ignorance, the names of these scientists and scholars have been expunged from the official list of Academy members.

17. Among those arrested in 1930–31 were a large group of plague and tularemia specialists. I came across this tragic story when working on a feature article about the victory over tularemia. A summary account of the fate of the arrested microbiologists may be found in my books (in Russian)

On the Track of Those Retreating (Moscow, 1963) and *The Map of Human Suffering* (Moscow, 1971).

18. Leningrad State Archives of the October Revolution and Socialist Construction, *VIR collection 9708,* file 520, folio 12.

19. Letter to the author dated April 14, 1975.

20. This was destroyed, together with four other scientific manuscripts. The fact was made known in a letter of September 4, 1965, from Semichastny, chairman of the KGB, to Kirillin, vice-president of the Academy of Sciences, ref. M 2008-C.

21. *Twenty Years of the Tashkent Medical Institute Named for V. M. Molotov* (Tashkent, 1939); article by Professor M. I. Slonim, pp. 22–23.

22. Ibid.

23. Ibid.

24. Archives of the Agricultural Academy, section 805, schedule 102, file 16 (section 1, file 198). Stenographic report of Presidium conference, May 29, 1931: report by Bursky.

25. Many *sharashki* (cf. p. 15 above) were in operation during the war for the purpose of devising new types of weaponry. The best known was that directed by the aircraft designer Tupolev, where a large number of scientists with higher degrees worked on the development of fighter planes. For the *sharashka* system see also Chapter 7.

26. Subsequently, on NKVD (KGB) orders, Shundenko prepared the case which led to Vavilov's arrest, and he spent the second half of his life as an agent of the security police.

27. Leningrad State Archives of the October Revolution, *VIR* collection 9708, file 1377, folios 15–16, 23.

28. The "party thousands" (*parttysyachniki*) were young Communists of proletarian or peasant origin who, on the party's recommendation, were directed into a scientific career regardless of their qualifications or general education.

29. The Higher Examinations Board, under the Council of Ministers of the USSR, has the task of confirming recommendations from the Academic Councils of universities and similar bodies as regards the conferring of higher degrees.

30. D. Gvishiani, "The Social Role of Science and Scientific Policy": lecture delivered at a symposium on the "Management, Planning, and Organization of Scientific and Technical Research," Moscow, May 1968, p. 4. As we have seen, in 1977 the total number of Soviet scientists was officially given as over 1,200,000.

NOTES TO CHAPTER 2

1. As part of his academic duties, the great scholar Mikhail Lomonosov had to write odes glorifying the Empress Catherine and to organize firework displays at palace receptions.

2. Skvortsov-Stepanov never actually occupied the academic chair, as he died in 1928 immediately after the "election."

3. "The best way to get rid of specialists is to replace them with ciphers. . . . Everyone knows that you can call in one of these nonentities, tell him to 'study the nature of man' and he will get to work and do so, without coming to any dangerous conclusions. A real chemist never knows when to stop, he keeps wanting to follow things through and achieve results; but one of

the biddable kind, after reaching a certain point, will wisely turn tail and see to it that others do likewise." M. E. Saltykov-Shchedrin, *Complete Works*, Vol. 7, p. 317.
4. This is one of the oldest kinds of test. For example, *Pravda* of January 28, 1937, at the height of the Stalin purge, published an appeal reading: "We demand that ruthless justice be meted out to the vile betrayers of our great country." The signatories were an elite group of academicians: A. Bakh (biochemist), B. Keller (plant breeder), N. Vavilov (geneticist), N. Gorbunov (secretary to the Academy), I. Gubkin (geologist and oil expert), E. Pavlovsky (parasitologist), A. Speransky (physiologist), N. Obraztsov (locomotive construction expert), P. Zdrodovsky (epidemiologist), and M. Lavrentyev (physicist).
5. B. N. Volgin, *Molodezh i nauka* (Youth and Science) (Moscow, 1971), p. 38.
6. Personal communication from A. A. Inozemtsev, vice-chairman of the Committee on Inventions and Discoveries.
7. Aleksandr Vasilyevich Sidorenko, Minister of Geology, was more successful. During his period of office he became a corresponding member (1963) and a full member (1966) of the Academy of Sciences.

NOTES TO CHAPTER 3

1. *Pravda*, October 8, 1975: speech by L. I. Brezhnev on the 250th anniversary of the Academy of Sciences of the USSR.
2. I. R. Petrov (1897–1972), head of the department of pathological physiology at the Leningrad Academy of Military Medicine. For further information about him see my book (in Russian) *The Way to the Heart* (Moscow, 1960).
3. *Meditsinskaya Gazeta*, August 9, 1968, p. 3.
4. "What Will Help the Academic Scientist," *Literaturnaya Gazeta*, No. 12, March 24, 1976.
5. Mark Popovsky, "The Hunt for a Toxin," in *Pravda*, February 1, 1972.
6. Ovchinnikov's staff declare, half-jokingly, that their laboratories are getting more and more like the "military colonies" established in Russia in the early nineteenth century, when soldiers who were obliged to serve twenty-five years in the Army were also made to perform forced agricultural labor.
7. A few theoreticians, especially physicists, form an exception to this rule.
8. This story was told me in December 1975 by a member of the Yoffe Physicotechnical Institute in Leningrad.
9. The Soviet leaders were twice captivated by the idea that chemicalization (as formerly electrification) would solve all the country's problems. In 1933–1934 the inspiration came from A. N. Bakh, and in the 1950s from Khrushchev's adviser, Academician N. N. Semenov.
10. I take this opportunity of expressing my gratitude to Dr. Selye for sending me his fine book *Stress Without Distress*, which touches on problems of medicine, biology, and social psychology. Unfortunately nothing like it has appeared in the Soviet Union in the past fifty years.
11. B. N. Volgin, *Molodezh i nauka* (Youth and Science) (Moscow, 1971), p. 37.

NOTES TO CHAPTER 4

1. D. Gvishiani, *The Social Role of Science and Scientific Politics* (Moscow, 1968), p. 31.

2. In thus setting twice as many assistants to work as the project required, the director was putting into practice the theory of a "scientific reserve," which we have already encountered in Chapter 2.

3. My article on Boshyan, "Sources of Life," was published in *Komsomolskaya Pravda*, May 10, 1950.

4. I was not allowed to expose the Boshyan fraud until twenty years later, in an article "Scientists and Others" in *Voprosy Literatury*, No. 2 of 1970.

5. The purpose of this department is to look after the health of top political and party officials. Its luxurious hospitals are supplied with the latest equipment and with drugs in short supply. Strictly closed to the public, there is one in every Republican and regional capital.

6. Professor Nikolai Ivanovich Khodukin (1886–1962), for many years director of the Republican Institute of Microbiology and Epidemiology at Tashkent, told me that the KGB once summoned him and asked: "Why don't you start an antidiphtheria campaign? Lots of people in the Republic suffer from diphtheria." "I didn't know that," replied the professor. "Well, *we* do," said the KGB man. "Start right away taking steps against diphtheria, but don't let anyone know it."

7. D. Mendeleyev, "What Kind of Academy Does Russia Need?" (1882), quoted from *Novy Mir*, No. 12, 1966, p. 189.

8. A small town near Tartu, where my wife and I spent the summers of 1974 and 1975.

NOTES TO CHAPTER 5

1. Znanie specializes in popular science and propaganda literature on scientific subjects.

2. Permission to publish my book was finally refused in February 1976.

3. Quoted from B. N. Volgin, *Molodezh i nauka* (Youth and Science) (Moscow, 1971).

4. See especially my *Panacea, Daughter of Aesculapius* (Moscow, 1973).

5. "Many times in the course of our journeys we were able to see for ourselves the meaning of scientific internationalism. If people knew your work and valued it in the slightest degree it was sufficient to send a letter beforehand and you were a welcome guest—they would go to the utmost lengths to help you, as if you were a close friend." N. I. Vavilov, *Five Continents* (Moscow, 1962).

6. Popova was wrong here. All letters from the USSR to foreign countries are censored.

7. I. V. Popova, letter to the author from Ramon, July 8, 1971.

8. Marquis de Custine, *Russia in 1839* (in Russian) (Moscow, 1930), p. 243.

9. M. E. Saltykov-Shchedrin, *Complete Works* (1965 ff.), Vol. 9, p. 433.

10. Pavlov's letter was published in full in *Na Literaturnom Postu*, No. 20, 1927, pp. 32–39.

11. "Maksim Gorky and Science," in *Izvestiya,* March 29, 1928.

12. See my *Panacea, Daughter of Aesculapius* (Moscow, 1973).

NOTES TO CHAPTER 6

1. See Chapter 1.

2. In 1966 *Prostor* published my documentary study "A Thousand Days in the Life of Academician Nikolai Vavilov," which was not included in any of my full-length books.

3. N. Bednoruk, "The Touchstone," in *Komsomolskaya Pravda,* February 13, 1974.

4. It is thought, however, that the removal of Mzhavanadze, First Secretary of the Georgian CP, and the transfer to Moscow, on promotion, of Mikhitdinov from Uzbekistan and Podgorny from the Ukraine, represent attempts by the Kremlin to combat extreme nationalism in the minority republics.

5. M. A. Leontovich is one of the outstanding figures in contemporary Soviet academic life, known for his pioneer work in connection with plasma and also for his crystalline honesty and moral integrity. At one time he and I appealed jointly to the Ministry of Internal Affairs to allow the political prisoners Yu. Galanskov and K. Bukovsky to receive better medical treatment. Leontovich is one of the few members of the Presidium of the Academy of Sciences whose mere presence suffices to restrain the archreactionaries. He voted stalwartly against the election of Lysenko's followers to the Academy, and has not signed any of the letters denouncing Academician Sakharov.

6. On an average, 22 percent of the population of the USSR has completed secondary school, while 4 percent have higher education.

7. At the beginning of 1976 an instruction in these terms was received by *Radio,* a journal concerned purely with radio electronics.

8. A series of such articles was published at the beginning of 1976 by the English-language *Moscow News.*

9. See Chapter 8.

10. Among the most important are G. I. Budker, director of the Institute of Nuclear Physics in the Science City at Novosibirsk; B. K. Vainshtein, director of the Institute of Crystallography of the Academy of Sciences of the USSR in Moscow; I. M. Khalatnikov, director of the Institute of Theoretical Physics; S. L. Mandelshtam, director of the Institute of Spectrography near Moscow; G. M. Frank, director of the Institute of Biophysics; and full members and corresponding members of the Academy of Sciences of the USSR such as V. L. Ginzburg, Yu. B. Khariton, M. A. Markov, L. B. Zeldovich, F. L. Shapiro, E. S. Fradkin, L. B. Okun, and others.

11. "The Position of Biological Studies," verbatim report of the session of the Lenin Academy of Agricultural Sciences, July 31–August 7, 1948 (Moscow: OGIZ-Selkhozgiz, 1948), p. 130.

12. Ibid., p. 524.

NOTES TO CHAPTER 7

1. See Chapter 1.
2. Norbert Wiener, *The Human Use of Human Beings: Cybernetics and Society* (London, 1954), p. 122.
3. Yuli Krelin, unpublished essay commissioned by the journal *Znaniye-Sila*.
4. *Literaturnaya Gazeta*, February 19, 1969.
5. I. V. Chernov and A. I. Shcherbakov, "Sociological Questions of the Organization of Labor," in *The Organization and Effectiveness of Scientific Research* (in Russian), 2nd ed. (Novosibirsk, 1967), p. 125.
6. Yuri Galanskov died in the camp at Potma, aged thirty-three, in the fall of 1972.
7. See, in Chapter 1, the story of the arrest of microbiologists working on tularemia (Gaisky, Elbert, and others).
8. A zealous proponent of this thesis is mathematician A. D. Aleksandrov, an academician from Novosibirsk-Akademgorodok. A detailed exposition of his theory that morality varies with the quantity of information available can be found in the collective work *Nauka i nravstvennost* (Science and Morality) (Moscow, 1971), pp. 26–73.

NOTES TO CHAPTER 8

1. Great Soviet Encyclopedia, 3rd ed., 1974, Vol. 18, p. 144, s.v. *nravstvennost* ("morality," corresponding to German *Sittlichkeit*. –Trans.).
2. The idea that a scientist is a mere instrument in the hands of the state, with no responsibility of his own, has often been proclaimed by top Soviet leaders. M. I. Kalinin, chairman of the Central Executive Committee and later chairman of the Presidium of the Supreme Soviet, wrote that the scientist's business was "not to invent what he likes, but what our socialist society needs"; and again: "When you carry out some piece of work it must be based not on your own inspiration but on the overall plan, the design presented to you." Speech at Higher Technical College, March 16, 1930; see *On Communist Education* (in Russian) (Moscow, 1956), p. 75.
   Addressing senior teachers in the Kremlin on May 17, 1938, I. V. Stalin declared that "progressive science" was the kind of science which "does not shut itself off from the people, does not keep aloof from the people, but is ready to serve the people, to transmit to the people all the conquests of science, to *minister to the people's needs not under compulsion but voluntarily and with joy.*" I. V. Stalin, *Works* (in Russian) (Stanford, Calif., 1967), Vol. 14, p. 275 (emphasis added). The word "people" should not mislead here: the leader of victorious Bolshevism always used it to mean "the state."
3. I wrote in my diary on February 24, 1971: "To praise a scholar, in a biographical work, for rendering service to his fatherland now seems to me as absurd as to praise the rain for falling opportunely. It is also stupid to blame him for not producing this or that. All we can and should do is to explain what intellectual riches a man is or was possessed of, and what use he made of them in accordance with his character, conscience, and external circumstances. . . . Alas, in my books I have not always escaped the 'taint of the times.' "

4. Maria Levitanus, letter to the author from Tashkent, March 16, 1973.

5. Iraida Popova, letter to the author from Ramon, July 8, 1971.

6. *Komsomolskaya Pravda*, May 11, 1966, p. 2. The letter to the editor under the title "Baikal Awaits" was signed by such eminent academicians as Artsimovich, Berg, Gerasimov, Zhukov, Zeldovich, Kapitsa, Knunyants, Kondratyev, Nikolsky, Trofimuk, Petryanov, and Emanuel. Their predictions came true: many cubic kilometers of water have been polluted by the plant in question, which—in addition to not paying its way economically—is still destroying flora and fauna of the utmost value.

7. Nothing is new under the Russian sun! In 1786 Sir Samuel Bentham, a naval architect and engineer in the service of Catherine the Great, erected on Potemkin's estate near Krichev a shipyard with a central structure from which the overseers, themselves hidden, could watch everything the workmen were doing. The idea was taken up by Samuel's brother Jeremy, the reformer, who defended it on theoretical grounds and proposed to build schools, prisons, hospitals, and factories on the same lines: the device was called the "Panopticon." In Britain it was rejected; but in Russia for the past two hundred years there has been no shortage of plans for the construction and perfection of Panopticons of all kinds.

8. This official viewpoint is also reflected in the latest edition of the Great Soviet Encyclopedia; see Vol. 15, p. 84, s.v. Lysenko, T. D.

9. See Chapter 7.

NOTES TO CHAPTER 9

1. In September 1975 the New York Cancer Research Institute honored Garri Izraelevich Abelev as one of the first fifteen immunologists in the world by presenting him with a medal for outstanding studies of the immunology of tumoral growths. They had much difficulty in handing over the medal, however, as the Soviet authorities would not allow Abelev to leave the country. The award was finally conferred in Moscow in a very modest ceremony in May 1976.

2. Scientists who, by virtue of their talents and prestige, have to some extent escaped control in the past are I. P. Pavlov, geochemist V. I. Vernadsky, and physicists I. E. Tamm and P. L. Kapitsa. In later times the authorities have been somewhat restricted in their control over Academician A. D. Sakharov. But for scientists of less eminence a show of independence usually means a dangerous and almost hopeless conflict with the regime.

3. "Saltomutations and Metamorphoses," *Bulletin of the Moscow Association of Natural Scientists*, Vol. 59, No. 4, (1954), pp. 67–79.

4. Ibid.

5. The vaccine was used on several thousand youths undergoing preconscription military training at the town of Gomel. All those inoculated suffered severe illness, and several dozen nearly died. A special conference of the party's Regional Committee was called, and the chief medical officer of the epidemiological center was dispatched to Moscow, where he declared that in thirty years' experience he had never known a drug that gave such appalling results.

6. G. Getting, *Meeting with Albert Schweitzer* (in Russian) (Moscow, 1967), p. 98.

7. A. A. Lyubishchev, "What Should We Be Like? My Message to Youth."

Interview given to R. E. Romanovsky on May 25, 1956. Original in Lyubish-chev family archives, Leningrad.

8. E. A. and G. I. Abelev, "Ethics as an Element in the Organization of Scientific Work." Manuscript, 1972.

9. The gifted biophysicist S. Kovalev (born 1938), author of sixty-five sci-entific works, was accused of circulating the underground *Chronicle of Cur-rent Events* and in December 1975 was sentenced to five years in a strict-regime camp.

10. "Statistically, I suppose slightly more scientists are in religious terms unbelievers, compared with the rest of the intellectual world—though there are plenty who are religious, and that seems to be increasingly so among the young." C. P. Snow, *The Two Cultures and the Scientific Revolution* (London: Cambridge University Press, The Rede Lecture, 1959), p. 9.

11. Decree No. 1067, approved by the Council of Ministers of the USSR on December 29, 1975. See *Bulletin of the Ministry of Higher and Secondary Technical Education,* No. 4, 1976, p. 15.

AFTERWORD:
About the Author

Mark Popovsky was born at Odessa in the Ukraine on July 8, 1922. After serving in World War II as a medical officer, he received a degree in philology at Moscow University and devoted himself to literature and journalism. He became a member of the Journalists' Union in 1957 and of the Soviet Writers' Union in 1961. By March 1977, having published seventeen books, he resigned from the Writers' Union in protest against the ill treatment of his fellow-writers. He himself was victimized on account of his frank political statements and for disseminating copies of prohibited books. By an order of February 1976 all publishers and editors in the USSR were forbidden to publish his work. In June of that year he addressed an open letter to the Sixth Congress of Soviet Writers. In 1977 he left the USSR permanently.

Mark Popovsky has written numerous studies, essays, and articles about scientists, which have appeared both in periodicals and in book form. These include biographies and the documented history of scientific investigations. He is especially interested in problems of scientific morality. Among his most important books in Russian are *The Way to the Heart* (1960), *The Story of Dr. Haffkine* (1963), *The Thousand Days of Academician Nikolai Vavilov* (1966), *The Map of Human Suffering* (1971), and *Panacea, Daughter of Aesculapius* (1973).

Popovsky's books combine rigorous accuracy with emotional involvement: as a researcher and critic he makes no secret of his feelings of love or hatred for those about whom he writes. Some of his favorite works have not seen the light of day in the Soviet

Union. Thus his book *Why a Scientist Needs a Conscience* (1975) was officially banned, as was the present work, and two large biographies: *The Misfortune and Guilt of Academician Nikolai Vavilov* and *The Life and Work of Professor Voyno-Yasenetsky*.